Creating Emotionally Intelligent Workspaces

Emotions in the workplace have until recently been seen simply as a distraction. We often think of work as rational, logical and non-emotional. But organisations are waking up to the key role of emotions and affect at work. Emotions influence how we make decisions, how we relate with one another and how we make sense of our surroundings.

Whilst organisations are slowly embracing the pivotal role of emotions, designers and managers of workplaces have been struggling to keep up. New insights from hard sciences such as neuropsychology are presenting a radically different interpretation of emotions. Yet workplace designers and facilities managers still rely on measuring non-specific states such as satisfaction and stress.

In this book we attempt to capture modern-day interpretations of emotion, looking at emotion in terms of transactions and processes rather than simple cause and effect. We entertain the idea of an 'emotionally intelligent building' as an alternative to the much-hyped intelligent building. The assertion is that we should create environments that are emotionally intelligent. Rather than focusing on the aptitudes or shortcomings of individuals at work, we should place closer attention on the office environment. It's not that we are emotionally disabled – it's the environment that disables us! The ability of you and me to interpret, control and express emotions may not simply be a result of our own make-up. A radically different outlook considers how our workspace and workplace debilitates or enables our emotional understanding.

In the modern workplace there are many innovations that can undermine our emotional intelligence, such as poorly implemented hot-desking or lean environments. Contrariwise there are key innovations such as Activity Based Working (ABW) that have the potential to enhance our emotional state. Through a series of unique case studies from around the world, we investigate key concepts that can be used by designers and facilities managers alike. No longer should designers be asked to incorporate emotional elements as intangible un-costed 'add-ons'. This book provides a shot in the arm for workplace design professionals, pointing to a new way of thinking based on the emotional intelligence of the workplace.

Edward Finch is a freelance 'thought leadership' author. He obtained his PhD from the University of Reading in 1989 and was Professor in Facilities Management at Salford University (2008–2011). He acted as Editor-in-Chief of the academic journal *Facilities* for 15 years. His writing challenges technological myopia in the workplace, suggesting that we also need to understand what it means to be human.

Guillermo Aranda-Mena is Associate Professor at RMIT University in Melbourne, Australia, and Visiting Professor of Architecture at Politecnico di Milano, UNESCO Chair in Mantua. In 2003 he completed his PhD at the University of Reading and has since collaborated with numerous universities around the world. He continues to travel widely and divides his time between Singapore, Hong Kong, Mexico, Italy and Melbourne, where he currently resides with his partner Dee and son Memo.

Creating Emotionally Intelligent Workspaces

A Design Guide to Office Chemistry

Edward Finch and Guillermo Aranda-Mena

First published 2020
by Routledge
2 Park Square, Milton Park, Abingdon, Oxon OX14 4RN

and by Routledge
52 Vanderbilt Avenue, New York, NY 10017

Routledge is an imprint of the Taylor & Francis Group, an informa business

© 2020 Edward Finch & Guillermo Aranda-Mena

The right of Edward Finch & Guillermo Aranda-Mena to be identified as authors of this work has been asserted by them in accordance with sections 77 and 78 of the Copyright, Designs and Patents Act 1988.

All rights reserved. No part of this book may be reprinted or reproduced or utilised in any form or by any electronic, mechanical, or other means, now known or hereafter invented, including photocopying and recording, or in any information storage or retrieval system, without permission in writing from the publishers.

Trademark notice: Product or corporate names may be trademarks or registered trademarks, and are used only for identification and explanation without intent to infringe.

British Library Cataloguing-in-Publication Data
A catalogue record for this book is available from the British Library

Library of Congress Cataloging-in-Publication Data
Names: Finch, Edward, author. | Aranda-Mena, Guillermo, author.
Title: Creating emotionally intelligent workspaces : a design guide to office chemistry / Edward Finch & Guillermo Aranda-Mena.
Description: Abingdon, Oxon ; New York, NY : Routledge, 2019. |
Identifiers: LCCN 2019028273 (print) | LCCN 2019028274 (ebook) |
ISBN 9781138602465 (hbk) | ISBN 9781138602472 (pbk) |
ISBN 9780429469558 (ebk) | ISBN 9780429891113 (adobe pdf) |
ISBN 9780429891090 (mobi) | ISBN 9780429891106 (epub)
Subjects: LCSH: Office layout–Psychological aspects. |
Work environment–Psychological aspects. | Emotional intelligence.
Classification: LCC HF5547.2 .F56 2019 (print) | LCC HF5547.2 (ebook) |
DDC 658.3/8–dc23
LC record available at https://lccn.loc.gov/2019028273
LC ebook record available at https://lccn.loc.gov/2019028274

ISBN: 978-1-138-60246-5 (hbk)
ISBN: 978-1-138-60247-2 (pbk)
ISBN: 978-0-429-46955-8 (ebk)

Typeset in Goudy
by Wearset Ltd, Boldon, Tyne and Wear

Contents

List of figures — viii
List of tables — ix
Foreword — x

1 **Introduction** — 1

   Leveraging the power of emotion 2
   What do we mean by emotional intelligence? 3
   Feelings or emotions? 6
   From 'know how' to 'know who' 6
   Summary 8

2 **Understanding emotions in the workplace** — 9

   The ABC of the human mind 9
   What is an emotion? 10
   Emotion in stages 10
   Types of emotion 11
   Positive psychology 13
   Emotions in practice 15
   Emotion, mood and temperament 15
   Emotion and the environment 17
   Summary 19

3 **Can a workspace motivate us?** — 21

   Form follows function 21
   Form follows emotion 22
   Evolution of the office 24
   What makes people tick? 25
   Maslow's hierarchy of needs 25
      Physiological needs 26
      Safety needs 27
      Affiliation and belongingness needs 28

vi Contents

        Status and esteem needs 29
        Self-actualisation 29
    Criticisms of Maslow's hierarchy of needs 30
    ERG theory 30
    Summary 31

4 The many faces of the intelligent building   32

    The death of permanence 32
    Layers within layers 33
    The emergence of the intelligent building (IB) 34
        The high-tech building 34
        The flexible building 36
        The interoperable building 37
        The green building 37
    The arrival of the 'sentient building' 38
    On collision 39
    Emotion detection 40
    Summary 40

5 The emotionally intelligent building   42

    Defining emotional intelligence 44
    Can a workspace have emotional intelligence? 45
        Ability model 46
        Mixed models 46
    Defining the emotionally intelligent workplace 47
    Making sense of it all 49
    The three dimensions of workspaces 50
    The green bus 51
    Putting the pieces together 52
    Emotional mechanisms 52
    Creating the emotional palette 54
    An emotionally intelligent work setting 55
    Summary 56

6 Emotion and the instrumental workspace   58

    Understanding the instrumental perspective 58
    The non-territorial office 59
    Laying down the law 60
    Fitting the person to the workplace 61
    User-centred design 62
    Nomad or vagrant? 63
    Replacing physical walls with psychological walls 63
    The emergence of Activity Based Working (ABW) 65

The truly territorial office 66
Behaviour change 68
Nudge 69
Summary 70

7   **Emotion and the aesthetic workspace**   72

Being there 72
Aesthetics and the senses 74
Aesthetics and emotions 74
Biophilia and biomorphic design 75
Fractals 77
Neuroaesthetics and neuroarchitecture 78
Summary 80

8   **Emotion and the symbolic workspace**   83

What is a symbol? 83
Symbolism and the organisation 84
The production of space 84
Workplace identity 86
Liminal space 87
Investigating a 'crime scene' 88
Enclosures 89
The IKEA effect 91
Our biological Wi-Fi 92
High-trust culture 92
Temperament 93
Summary 94

9   **Conclusions**   96

Reflections 98

Index   99

Figures

1.1	Office chemistry	2
1.2	A front-of-house corporate setting	5
1.3	Creation of incubator space in a remodelled church	7
2.1	The stages of emotional response	11
2.2	Emotions and the brain	12
2.3	Emotion, mood and temperament	16
2.4	The affective qualities of space	18
3.1	Form follows emotion	23
3.2	The elevator model of Maslow's human needs	27
4.1	The emergence of the intelligent building	34
4.2	The ABC path of office evolution	38
5.1	Space efficiency versus delight	42
5.2	The emotionally intelligent building (EQ) versus the intelligent building (IQ)	44
5.3	The dimensions of an artefact (instrumental, aesthetic and symbolic)	53
5.4	The different mechanisms influencing instrumentality, aesthetics and symbolism	53
5.5	Designing with the emotional palette	55
6.1	Open plan design office	64
6.2	Determinants of anger in the territorial workplace	67
7.1	Creating 'naturalness' in the work environment	76
7.2	Fractal geometry	77
8.1	Looking at space through a different lens (Lefebvre's triad and the social production of space)	85
8.2	Illustration of liminal space with corporate memorabilia	88
8.3	Investigating a 'crime scene'	89
8.4	Use of partitions in an open plan office	90

Tables

5.1 Contrasting assumptions between the intelligent workspace (IQ) and the
    emotionally intelligent workspace (EQ)   48
5.2 Workspace characteristics expressed as instrumental, aesthetic and
    instrumental dimensions   54

Foreword

Both of us have the pleasure of writing the foreword for this book. Claudia met Edward towards the end of 1999, at an EDRA (Environmental Design Research Association) Congress in Orlando. At that time, both he and Claudia were part of a global research team led by Prof. Dr. Wolfgang Preiser, the leading authority in post-occupancy evaluation. Izabel met Edward many years later, whilst practicing workplace strategy and co-teaching a Corporate Workplace Management class with Claudia at the University of São Paulo (USP).

This co-authored book is the culmination of a two-year collaborative venture between Guillermo Aranda-Mena in Australia and Edward Finch in the UK which has taken them on a surprising journey.

We all share an unwavering passion for understanding the workplace and its effects on people and their relationship with the physical space. With the advent of the new millennia, technology has emerged that has put an end to the office as we knew it – concepts of mobility, non-territorial offices and smart buildings to name a few. But this never took our focus away from the asset that matters most in the success of all organisations: the people.

When organisations need to refocus their culture and processes and implement new ways of working or adopt new technologies, the workplace needs to be reframed beyond comfort, ergonomics and functionality. There is something more subjective, more subliminal, that brings to the surface people's sense of belonging, reinforcing their identity and creating the emotional stimuli necessary to be productive and fulfilled at work – even with the technological apparatus of our times.

This book bridges the logical and the emotional sides of people at work, raising a new challenge: for designers to conceive of the office beyond its physical attributes. It challenges facilities managers and designers to create the 'chemistry' of what makes people 'click' with a certain place and want to become a part of it, turning it into a place of choice. Spaces need to become responsive to the needs of an emotionally diverse workforce.

From our extensive experience working with corporate and public-sector clients, the value of the workspace cannot be considered simply as a cost for an organisation. Its value should focus on the ability of the workplace to promote social engagement within an organisational culture – whilst meeting an organisation's business objectives. In this context, the book presents two important realities that should make us pause and reflect: (1) that the physical environment can be intentionally used to nurture fruitful emotions; and (2) that human relations at work are changing to a point where people cannot be 'owned' by organisations any longer. The challenge becomes the creation of workspaces where people can make a grassroots contribution to defining their own organisations.

It goes beyond working anytime, anywhere – it is about what people want to be part of and where they want to belong.

These arguments are substantiated by neuroscience and the evidence of future trends. Such insights will inform design development as a cooperative process involving workers as active participants, rather than passive users. This evidence base will change how we approach workspace design. The approach by Finch and Aranda-Mena concerning the emotionally intelligent workspace brings these issues to the forefront and represents a seminal contribution to workplace design.

The theme of emotion unravels the soul of corporations, creating new, more humane spatial relationships. The emotionally intelligent workspace reveals a much more exciting and challenging future for all of us.

Claudia Andrade and Izabel Barros

Claudia Andrade is a Brazilian architect, a specialist in workplace strategy and organisation effectiveness, with a PhD in Workplace Performance Evaluation from the University of São Paulo. Izabel Barros is a Brazilian engineer and designer, a specialist in workplace strategy and organisation efficiency, with a PhD in Innovation and Strategy from the Institute of Design at the Illinois Institute of Technology in Chicago. They have years of experience implementing new ways of working in the most varied organisations, industries, countries and cultures. Their extensive involvement with organisations across the world, developing workplace strategy, workplace culture, experience design and change management, has contributed to the organisational success of many global clients.

1 Introduction

The modern-day office depends upon human interaction and – more importantly – engagement. Communication technology might allow us to work from anywhere but face-to-face interaction still plays a vital part in organisations. Presented with so many options about where we work, how do we make the office the place of choice?

Discussions about 'relationship chemistry' bombard us daily in popular media. Relationship chemistry describes a bundle of emotions that two people get when they have a special connection. It is the impulse that makes you feel, "I would like to see this person again." How often do office environments give us the same feeling – that we are able to 'click' with it? Can designers and facilities managers play some role in creating this elusive chemistry? This book explores the untapped potential of workspaces that we occupy, spaces in which we spend an increasing proportion of our waking day.

Good chemistry is not just about person-to-person relationships between two people. The growing importance of collaborative working has redoubled our attempts to understand exactly what makes teams tick. Group interaction is only possible if the chemistry is right. One leading organisation, Deloitte, has developed its own business team entitled Business Chemistry with the express intent of improving the art of relationships. And when it comes to solitary undisturbed working environments, the chemistry between the office worker and their environment becomes pivotal. The emergence of 'deep working' (Newport, 2016) in people's work routine necessitates a very different workspace. How do we get the chemistry right and meet these very different workspace needs?

Some of the characteristics of good chemistry in a relationship include non-judgement, mystery, attraction, mutual trust and effortless communication. How often do we encounter this type of relationship in our work environment? Wouldn't it be special if the office environment itself encouraged a sense of mystery, attraction, mutual trust and effortless communication? Interior lighting, space planning, user control, colour schemes and thermal comfort are just some of the design considerations that have been shown to influence individual and group behaviour. More often than not, studies focus on the 'cognitive' and 'behavioural' interaction between office workers and their environment. But what about emotion (and the related concepts of affect or feeling)? A growing number of organisations now recognise just how much they have neglected the 'A' in the ABC of psychology (affect, behaviour and cognition). In a world where loyalty and belonging have become a rare commodity, organisations are reaching out for ways to create a 'trust-based' culture. Millennials, unlike their predecessors, have grown up in an era where the 'contract' is king. They have become accustomed to fixed-length contracts and are more inclined to switch between employers. The office environment has become one of the few ways to attract and retain this talent.

Figure 1.1 Office chemistry.

Leveraging the power of emotion

So what exactly is an 'emotionally intelligent' workspace? In some ways it runs counter to the idea of an 'intelligent building'. Whilst the intelligent building attempts to leverage the power of technology, the emotionally intelligent building attempts to leverage the power of human emotion. In the business world, personal intelligence is no longer seen as the main indicator of job performance. A much better predictor appears to be 'emotional intelligence':

> Emotional intelligence can be defined as the ability to monitor one's own and other people's emotions, to discriminate between different emotions and label them appropriately, and to use emotional information to guide thinking and behavior.
> (Colman 2015)

What part does technology play in all of this? One might imagine a science-fiction scenario whereby the building itself is able to monitor the emotions of its occupants and perhaps use this information to guide its own behaviour. There are indeed examples discussed in this book that illustrate this possibility and the ethical implications. But building technology has not always been supportive of human emotion. Indeed, what has been considered as building intelligence can undermine emotional intelligence. In the pursuit of a seamless network of computer and human interfaces, we can end up with an office environment that is entirely process driven. The office as a knowledge exchange is thus measured only in terms of connectedness.

At the other extreme, prestige office buildings can sell themselves in terms of luxury, delight and novelty. Such buildings have proven to be effective at enticing new employees. Rather than using the appeal of new technology, such buildings rely on the experiential aspects of natural light, space and furnishings. Undoubtedly these buildings provide the opportunity to engage with people's emotions. But it seems that this is often arrived at by chance or design hunches. What we want to do in this book is create a framework that uses an evidence-based approach to emotional intelligence. In other words, is it possible to design and operate a building that is sensitive to the emotional needs of its occupants? This is not about pandering to whims. It is about supporting the relationship chemistry that is becoming so important to successful organisations.

What do we mean by emotional intelligence?

Intelligence is overrated. That is what the evidence has shown us. Whilst there is much debate over what intelligence actually is, we consistently find that, however you measure it, it is a very poor predictor of personal success and happiness. This applies whether we are looking at our effectiveness in business or in our personal relationships. Yet there remains an unquestioning reverence for the technology-laden 'intelligent building' or the 'smart building'.

In just the same way that we measure personal intelligence using IQ, practitioners have attempted to assess a building's IQ. Early formulations of the intelligent building in the 1980s and '90s saw increasingly complex but dedicated systems for energy management, lighting control, air-conditioning systems and other self-contained building technologies. Moving into the new millennium we witnessed the complete integration of these systems using complex building management systems. Whilst these advances improved the efficiency of the building, their impact on building users was more subtle. Undoubtedly the intelligent building has enabled office users to engage almost seamlessly with computers and other workers. The advent of universal Wi-Fi in the workplace has allowed entire workforces to become truly mobile. It is now possible to have an uninterrupted transition between the office, home or anywhere else. Furthermore, the ability to track and monitor user behaviour gives us an unprecedented understanding of how a building performs.

Despite these new advances, the modern office faces more challenges than ever. Some commentators have predicted the demise of the office. Office designers and facilities managers have had to do a rethink. What is it that an office provides that a home working environment seems unable to fulfil? Organisations demand workplaces that are more than comfortable desks set up with the latest kit. Offices need to become indispensable parts of an organisation: places of choice where trust and loyalty are nurtured.

Perhaps we have exhausted the idea of the intelligent building? Or perhaps in our pursuit of building intelligence we have neglected something. In this book we argue that the modern workplace needs an entirely different kind of intelligence: emotional intelligence. In just the same way that emotional intelligence (EQ) has proven to be a much better predictor of success at work and in people's personal life, the equivalent measure could be used to identify emotionally intelligent buildings. We might try to assess a building's capacity to match the emotional needs of groups and individuals in the workplace.

4   Introduction

But let's not get hung up on bean counting. More than anything, we need to 'frame' the problem that is in front of us: how to harness human potential.

Is this about technology? Whether you are an architect, facilities manager or indeed any professional involved in leveraging human potential through the built environment, you are in the business of emotional intelligence in all its forms. Emotional intelligence can be thought of as the ability to:

- recognise, understand and manage our own emotions
- recognise, understand and influence the emotions of others.

We might have some difficulty imagining a building that can recognise and understand its own emotions (unless of course you are a science-fiction enthusiast). However, the second item, 'being able to understand and influence the emotions of others' clearly has relevance to workplace design. Isn't it something that architects and interior designers do all the time? Indeed, lighting, fixtures and fittings, colour schemes and soundscapes are devices that are routinely used to create ambience. But is this emotional intelligence? Is it enough to simply create the desired emotional setting? What level of user control should there be? What about influencing emotions in a shared environment? If a building cannot recognise emotions, can we say that it is capable of influencing our emotions?

Resorting to headphones

Zhi has always adopted an 'open-door' policy working as a senior partner in her organisation. Recently her organisation has moved to an open plan hot-desking work environment. She can no longer express her open-door policy because they have taken the door away and the walls as well. In fact, sometimes she would like to have a 'closed-door' policy, but without a door she has had to resort to using headphones.

Zhi's predicament is perhaps a rather simplistic illustration of how we use the physical environment to express emotions. Take away a key part of that setting and suddenly we feel rather exposed. Workplace designers have been busy removing fixtures and fittings that we rely on to convey emotions. The 'non-stick' environment eradicates 'emotional potential'. In our attempts to create the intelligent building, we thwart the emotionally intelligent building.

Throughout this book we examine the devices and interventions that might be used to enhance emotional intelligence. As much as possible, the analysis is through the 'emotionally intelligent' lens. Rather than parading a set of interesting possibilities, the book attempts to develop a framework to capture the emotional response of building users. It presents an 'emotionally intelligent' language to enable diverse design teams to share a common understanding.

Can a building really be emotional? To answer this question, let's first of all look at the modern-day office – an example of a complex and layered system. In order to work, it demands the close integration of a physical system, an IT infrastructure and people (the facilities management team). But the physical system is itself composed of different elements of varying lifecycles. Instead of just bricks and mortar, the contemporary workplace is comprised of 'shell, services and sets'. This layered approach identified by Duffy conveys how the modern office enables technological and organisational change:

Introduction 5

> Our basic argument is that there isn't any such thing as a building. A building properly conceived is several layers of longevity of built components.
>
> (Duffy 1990, p. 17)

Whilst the shell or building structure remains relatively unchanged, changes to building services typically occur every 10 to 15 years as fit-outs respond to the changing demands of clients. Rates of change are even more dramatic when we look at the furniture and fittings (sets) to be found in today's office. This innermost layer of the office 'kit of parts' represents a progressively large constituent, capable of meeting both human and IT demands. It is this soft, fluid layer that can undergo changes almost on a daily basis. This level of orchestration relies on the interventions of people – both users and, where necessary, the

Figure 1.2 A front-of-house corporate setting.

facilities management team. In reality, the modern office is far from being a static monolith: it is a living system. Adaptability is driven by human needs rather than technological needs. It is this capacity to accommodate changing human needs that defines the emotionally intelligent building.

Feelings or emotions?

Isn't the modern work environment about rational thought and behaviour? Don't emotions simply get in the way of rational decision-making? This has remained the longstanding belief of many organisations and management theorists. But a growing number of practitioners now acknowledge the pivotal contribution of emotions in organisations. There is an emerging realisation that the physical environment can be used to manage emotions and improve work outcomes.

If you are a psychologist, you'll often refer to feeling or emotion as an *affect*. It comes first in the ABC of psychology (affect, behaviour, cognition). It's a psycho-physiological construct. Put another way, it hits you in the heart as well as the mind. It's not just about feelings – we can actually measure responses. It affects our peripheral physiology (e.g. release of the stress hormone cortisol and heart rate), our actions (e.g. facial expressions and our fight or flight response) as well as our cognition (e.g. vigilance) (Bradley and Lang 1994).

Cognition fits much more easily with our 'rational' view of the organisation. It embraces processes such as knowledge acquisition, perception, attention and memory. Notice that we can recognise each of these processes as computing terms. It is perhaps no surprise that cognition is the one human capability that computers and AI are increasingly replacing.

We often use the words 'temperament', 'mood' and 'emotion' interchangeably. By clarifying the differences between each of these three concepts we expose some important insights. Psychologists generally agree that temperament, mood and emotion can be mapped out on a temporal continuum starting off with short-lived emotions.

From 'know how' to 'know who'

There's something afoot that is transforming the office market. Something so fundamental that it is permanently altering how we think about office real estate. It's a change that will move 'emotional intelligence' to the front of the line. The tidal change to which we are referring is 'social capital'. Just as we had got used to the idea of the 'knowledge economy', it seems that organisations now have to go one critical step further, a step that will give them a key advantage. Organisations are no longer preoccupied with 'owning' human capital (including their skills, knowledge and experience). What is much more important to them is having 'access' to human talent. Organisations wants to accumulate 'social capital' – the elusive threads that connect up human capital. It's no longer 'what you know' but 'who you know'. This change is also linked to new working practices and the emergence of the freelance worker and start-up or incubator space (such as that shown in Figure 1.3).

What's exciting is that the emergence of social capital places office design centre stage. For intelligent buildings conceived in the first wave of the knowledge economy, providing access to information (including mobile data) was the key challenge. Technology provided all the answers in the form of wireless communication and environmental control. Today it seems that Wi-Fi connectivity and environmental comfort is a seamless part of everyday

Figure 1.3 Creation of incubator space in a remodelled church.

life. But the modern office has to do more than that. For emerging organisations, the office provides the *context* where information is interpreted, combined and repurposed to produce new products and services. Knowledge is not simply acquired – it is transformed as part of an emerging social network involving partners, customers, suppliers and co-workers. Organisations are no longer in the business of *owning* human capital: access is what it's about. The organisations that can design a context for creative and collaborative relations will lead the pack. The physical office rather than the virtual office is where it's being played out.

Linked to the emergence of social capital is the practice of coworking. Organisations no longer need to exercise 'command and control' in the office. Remote working offers many advantages, including reduced real estate costs and reduced travel time for employees. But the practice of homeworking leads many people feeling isolated and socially adrift. This has led to the emergence of coworking spaces – "shared spaces where individuals do their own work but in the presence of others with the express purpose of being part of a community" (Garrett *et al.* 2017, p. 821). It provides a middle ground between the traditional office and working from home. It is commonly associated with unaffiliated freelance creative workers or contract workers. However, a burgeoning trend is evident in public- and private-sector organisations who recognise that coworking provides a more attractive alternative to homeworking. For the office designer, the practice of designing for a particular organisational 'culture' disappears. Instead, the challenge is to design an environment of coproduction involving an eclectic mix of individuals. Design becomes a cooperative process involving participants rather than users.

Summary

The office landscape is changing rapidly – new working practices are emerging that challenge all our assumptions. Nascent forms of workspace are appearing that have to satisfy a diverse range of individuals. No longer is it sufficient for designers to rely on a brief based on a homogenous organisational culture. We need to understand how design influences individuals in different ways. Diversity becomes both a challenge and an asset. In the next chapter we explore the science of emotions while avoiding faddish psychobabble. This paves the way for Chapter 3 where we examine how emotions interact with motivation – or what makes us 'tick' in the workplace.

References

Bradley, M.M. and Lang, P.J., 1994. Measuring emotion: The self-assessment manikin and the semantic differential. *Journal of Behavior Therapy and Experimental Psychiatry*, 25 (1), 49–59.

Colman, A.M., 2015. *A Dictionary of Psychology*. Fourth Edition. Oxford and New York: Oxford University Press.

Duffy, F., 1990. Measuring building performance. *Facilities*, 8 (5), 17–20.

Garrett, L.E., Spreitzer, G.M. and Bacevice, P.A., 2017. Co-constructing a sense of community at work: The emergence of community in coworking spaces. *Organization Studies*, 38 (6), 821–842.

Newport, C., 2016. *Deep Work: Rules for Focused Success in a Distracted World*. London: Hachette UK.

2 Understanding emotions in the workplace

In the introductory chapter we highlighted the organisational drivers that are demanding a reinvention of the workspace. Organisations are haemorrhaging talented employees simply because they failed to engage them at a deep-seated emotional level. The design and day-to-day operation of the workplace has the capacity to re-engage with employees. Indeed, many organisations are throwing significant resources at this problem. But a more in-depth understanding of workplace emotions suggests a radically different setting from today. In this chapter we examine the current understanding of emotions, as informed by current psychological and neurological research.

The ABC of the human mind

Psychologists talk about the 'ABC' of psychology – affect, behaviour and cognition. The word 'affect' is often used interchangeably with the words 'emotion' and 'mood'. Whilst it appears as the first item in the ABC sequence, it is the one pillar that has received the least attention in traditional management theory. What has attracted much greater attention is the 'C' of cognition. Cognition refers to the mental action or process of acquiring knowledge and understanding through thought, experience and the senses. It encompasses all those activities that we associate with 'knowledge management' and the 'rational' world of the office. We can also see direct analogies between the human brain and that of the computer (sensor, memory, nervous system/network) as a link in the information-processing chain. Looked at this way, the contribution of people in the workplace is simply as an extension of the connected digital brain of the organisation. Following closely in its path is the 'B' of 'behaviour' – a concept keenly familiar to designers – referring to the range of actions and mannerisms made by individuals. Behaviour, unlike cognition and emotion, is something we can observe directly. When it comes to space usage, we now have a number of tagging and tracking technologies that automate much of the process involved in post-occupancy evaluation (POE). But observation and measurement only take us so far, giving us only the answer to 'what' people do during a working day. As designers, we are left with the question as to 'why' people behave in a certain way. This can only be answered by addressing the 'A' or affect/emotion. Unlike behaviour, which is observable, emotion is subjective. Our emerging understanding from the science of emotion gives us compelling reasons to revisit how we design our office space.

What is an emotion?

We have all experienced intense emotion. Different surroundings certainly have the power to evoke these strong feelings. A poorly lit pedestrian walkway can make us feel threatened late at night; a spacious atrium can fill us with awe; or perhaps the discovery of certain artefacts can evoke strong feelings of nostalgia. These intense feelings are often accompanied by thoughts and physiological changes and can prompt us to behave in a certain way (such as a desire to flee from a threatening environment). Later in this chapter we will look at mood and temperament, two states that widen our discussion of the emotionally intelligent workspace. But for the moment, it's useful to pin down in a more scientific way exactly what we mean by emotion. All emotions appear to exist to enable us to function and survive in response to our internal and external environment. As such they are described as psychological adaptations. Carlson and Hadfield describe emotion as:

> a genetic and acquired motivational predisposition to respond experientially, physiologically, and behaviourally to certain internal and external variables.
>
> (1992, p. 5)

Another definition emphasises how emotions in more developed animals allow them to function beyond simple instinct:

> ... higher organisms experience a much wider variety of needs and goals, and instead of merely sensing environmental events, they interpret them. Interpretations of events that are particularly relevant to such needs lead to emotions. Emotions, in turn, motivate the organism to respond to its environment, but allow an adaptive flexibility of response that is not available to organisms that rely on instinct.
>
> (Ellsworth and Smith 1988, p. 272)

Another observation by Herbert Simon suggests that "a chief function of emotion is to interrupt and reorder processing priorities" (Simon 1967). As such, emotion has a pivotal role in changing the world that we see around us. In fact, recent evidence based on scientific experiments examining visual perception suggests that "emotions routinely affect how and what we see" (Zadra and Clore 2011, p. 676).

Emotion in stages

What happens when we experience an emotion? We can all think of our response to an extreme emotion like anger: our hair stands on end; our heart starts pumping quickly; our feelings are shown in our facial expressions; suddenly we focus on that one situation in front of us; and we then behave in a certain way. We also subjectively feel the emotion and perhaps have the ability to express it in words. We can say that emotion involves biological responses (physiological); information processes (cognitive or psychological response); and a behavioural response. An emotional response involves a number of underlying stages (Stanley and Burrows 2003, p. 7), also shown in Figure 2.1:

1. Detection of the event.
2. Change in arousal (preparing and orienting in order to respond).
3. Appraisal of the event in terms of significance (interpretation).

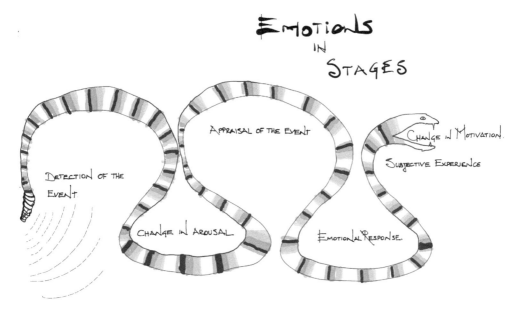

Figure 2.1 The stages of emotional response.
Source: Stanley and Burrows 2003, p. 7.

4 Emotional response.
5 Subjective experience of the emotion.
6 Change in motivation.

The 'arousal state' enables us to prepare for a response. It allows us to draw on our internal resources as required. The change in arousal state depends upon our baseline (basal) state prior to the event. For example, if we are already in a heightened state of anxiety, our response may be significantly amplified compared to a relaxed individual. Constant throughout the system is a feedback mechanism that attempts to restore balance to the individual (homeostasis).

Types of emotion

Our modern-day understanding of emotions has emerged from physiological, psychological and more recently neurobiological research. Perhaps the most revealing work about the nature of emotions has come from neurological work on deep brain stimulation on various animal types (Montag and Panksepp 2017), as shown in Figure 2.2. In these studies, three evolutionary 'passages' can be identified in the brain that influence emotion:

1 the reptilian (deep cortical) part that has evolved from an ancient evolutionary pathway;
2 the old-mammalian (limbic) that is also part of an ancient neural system;
3 the neo-mammalian (neo cortical) that constitutes a much more recent evolutionary development.

12 *Understanding emotions in the workplace*

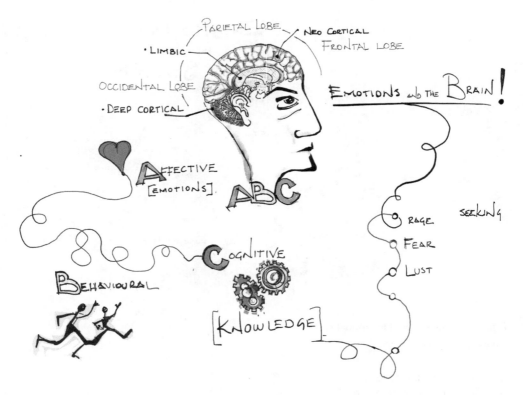

Figure 2.2 Emotions and the brain.

If we look first at the reptilian (deep cortical) part of the brain, we observe a system responsible for autonomic processes like breathing and heartbeat. Alongside these life-sustaining activities we can also locate ancient emotions that are identified as FEAR, LUST, RAGE and SEEKING.

Representing a further evolution of the brain (although still ancient in origin), the limbic system provides a further layer of sophistication and complexity to the emotions identified in the reptilian system. But perhaps more importantly, we see the inclusion of additional emotions that enable animals to function as part of a social system. These emotions include maternal CARE, social PLAY, and separation-distress or PANIC/SADNESS.

Finally, we have what might be described as the 'thinking cap' or the neo cortical region which is particularly well developed in humans. This area of the brain allows us to cognitively (knowingly) regulate emotions.

The way in which we respond to things around us will vary from person to person. It is these differences that give rise to our own personality. Such differences are genetically dictated by our primal systems (deep cortical and limbic) and the way in which they interact with our filtering system (neo cortical).

Does a knowledge of emotional anatomy allow us to make more informed decisions about the office environment? Evidence shows us that we are hardwired genetically, and through adaptation we use the same emotional infrastructure to those of our ancestors (and other living mammals). Evolutionary psychology suggests that natural selection has provided humans with many psychological adaptations, in just the same way that it generated anatomical and physiological adaptations. Whilst in early man such adaptations were used to secure food, find a partner or obtain shelter, these very same adaptations are used in modern-day life (including the office) to function effectively.

Perhaps the most interesting emotion amongst the positive emotions of SEEKING, LUST, CARE and PLAY is the SEEKING emotion. It is a general-purpose system that enables us to secure resources from the world around us. It can be described as our appetitive-exploratory-investigatory system. In the modern office environment, we see many devices that are used to encourage this SEEKING behaviour. 'Line of sight' in the open plan environment is one example that is used to encourage curiosity and exploration – a fundamental part of the SEEKING behaviour. Being able to see a colleague across an open plan space is more likely to encourage further interaction and possibly the securing of a useful resource. SEEKING is a general-purpose emotion that oversees 'motivational systems'. These motivational systems have dedicated need-state detectors that help us to maintain bodily states (homeostasis) such as energy (hunger) and water (thirst). But it is the SEEKING emotion that appears to bring together information from these need-state detectors. We can therefore see that many of the goal-directed behaviours in the office, such as eagerness and enthusiasm, are dependent on the SEEKING urge.

The SADNESS emotion appears to be closely linked to SEEKING. This reflects the importance of humans as a social animal. Whilst the positive emotion of SEEKING allows the development of social bonds and cooperation in the modern work environment, the PANIC/SADNESS circuit is typically accompanied by negative feelings associated with loneliness:

> When one is alone, whether mouse or man, it is simply more difficult to survive, and to get through life with an affectively positive state of mind.
> (Montag and Panksepp 2017, p. 3)

We typically try to reduce the circuit activity associated with SADNESS and increase the level of activation associated with SEEKING in order to increase our well-being and sense of companionship at work.

Positive psychology

Modern psychology has thrown open an entirely new window on the world of people and their interaction in the workplace. 'Positive psychology' is one area that has had a particular impact. This is a term coined by Seligman (then president of the American Psychological Association) who questioned why advances in modern psychology since 1947 were almost exclusively concerned with the treatment of mental illnesses (Seligman and Csikszentmihalyi 2000). He expressed the concern that "although psychology had become proficient at rescuing people from various mental illnesses, it had virtually no scientifically sound tools for helping people to reach their higher ground, to thrive and to flourish" (Seligman and Csikszentmihalyi 2000, p. 5). Both researchers suggested that we need to understand more about how optimism and hope affect health, what constitutes wisdom, and how talent and creativity come to fruition.

Connected with the idea of positive psychology, the importance of positive emotions has also received greater attention from recent studies. Foremost amongst these is the work of Fredrickson (2001, 2003) who has questioned why negative emotions have tended to dominate people's thinking on the subject of feeling and affect. She has suggested that negative emotions attract attention for the following reasons:

- Negative emotions can easily be coupled with particular behaviours, such as anger that creates an urge to attack, fear that creates an urge to escape and disgust that creates the urge to expel (expectorate).
- Negative emotions give rise to particular facial expressions that are universally understood. For example, we can readily identify angry, sad and fearful faces.
- Negative emotions often produce distinct physiological changes. For example, fear prompts an increasing blood flow in preparation for a flight response.

When we consider some of the positive emotions it is often difficult to discriminate between them. First, there is no clear linkage between an emotion and the behaviour that often follows from it (e.g. joy, amusement and serenity). Furthermore, these positive emotions are often indistinguishable in terms of facial expression. All of the positive emotions give rise to what is known as the Duchenne smile in which the lips become raised at the corner and the muscles are contracted around the eyes. Physiological changes are also less evident when positive emotions are experienced.

Why shine a light on positive emotions? From an evolutionary perspective the role of negative emotions is clear: they enabled our ancestors to survive in life-threatening situations – they are concerned with immediate survival. In contrast, positive emotions seem to allow personal growth and development. They provide a longer-term solution for resilience in the face of hard times. "Simply by experiencing positive emotions, our ancestors accrued more resources" (Fredrickson 2003, p. 332).

Fredrickson (2003) describes a fascinating experiment in which students were invited to watch a number of emotionally evocative film clips. For example, the emotion of 'joy' was elicited by watching a herd of playful penguins. The participants in the study were then assessed in terms of their ability to think broadly using global-local visual processing tasks. The experiment revealed that those who experienced positive emotions tended to see the 'big picture'. In contrast, those in negative or neutral states tended to confine themselves to narrowly focused detail. In other words, our emotional states affect how we see the world. This led Fredrickson to propose the 'broaden-and-build' theory which suggests that "positive emotions broaden an individual's momentary mindset, and by doing so helps to build enduring personal resources" (2003, p. 332). The resources to which she referred included the discovery of novel ideas, actions and social bonds.

Why do we need to create an office environment that stimulates positive emotion? After all, shouldn't we simply be eliminating negative emotions? Perhaps we should concern ourselves with "making miserable people feel less miserable" (*The New Era of Positive Psychology* 2012). Organisations are increasingly aware of the 'pull' of the work environment, both in terms of attracting top talent and minimising attrition rates. There is also a general acceptance that 'happy workers make productive workers'. But the evidence regarding positive emotions is even more compelling than we thought. In fact, Fredrickson's 'broader thought-action repertoire' (2001) suggest that positive emotions actually make us think better. When we experience things that produce positive feelings, they can significantly improve our creativity. Other experiments involving a group of physicians found that participants who were made to feel good by giving them sweets were able to solve problems related to a patient's liver problem in a more creative and integrative way (Estrada *et al.* 1997). It was found that the physicians were less inclined to anchor on a premature conclusion and were able to see the bigger picture.

It seems clear that emerging ideas in psychology and neuropsychology are providing the creators of workspaces with an entirely new lens on the world. We are no longer confined to traditional environment-behaviour research but instead we can start to look under the bonnet. Rather than examining immediate behavioural responses, we are beginning to understand how the physical environment furnishes us with personal resources and social resources. Instead of a preoccupation with negative emotions, we can also start to understand 'what makes life worth living'.

Emotions in practice

> It wasn't supposed to be like this. It started off as a simple entry in my electronic diary. Now it's a conference-call gone wrong. It took three attempts this morning to get a desk that had the right configuration and monitor connection for my laptop. That was 20 minutes in all. Now I've missed my call with my colleagues in Singapore. This situation is driving me insane.

Emotions can sometimes run high in the office. Perhaps for our conference call it is just a matter of sorting out the technology. But if we read between the lines, we see emotions or feelings are also at work. Many feel that emotion has no place in workplace environments. Its unpredictability does not sit well with our immutable entry in the diaries. We might also feel that it does not mesh with group working which seems to demand that we are always predictably 'bright-eyed and bushy tailed'.

The reality is, emotion in all its forms constitute one of our greatest human assets. In fact, it is this primaeval circuitry that enables us to respond to novel situations using a highly adaptive neural system. As noted by Payne and Cooper (2003, p. 10):

> The denial of emotional factors in the workplace is not realistic. The failure of the workplace that attempts to suppress emotion, and the realisation that a positive emotion may be beneficial to work outcomes, leads to both employee-assistance programs and teambuilding, and rewards-based work practices.

Despite its importance, we often do not know how emotion works. As workplace designers or facilities managers, possessing the tools that tap into this elusive thing we call 'emotion' has become more important than ever.

Emotion, mood and temperament

> Sasha nervously negotiated the revolving doors to her new workplace with a latté in hand. People always saw her as a 'glass half empty … or rather a coffee cup half empty' kind of person, and her anxieties about keeping her new job played heavily on her mind. There had been an atmosphere in the office over the last few months about possible redundancies because not enough work was coming in. She was always reminded of the mantra 'first in, first out'. Not being a morning kind of person, she rarely lifted her head to acknowledge the receptionist, who always looked up from her perch with a ready smile whatever the weather.

16 Understanding emotions in the workplace

> Entering into the open plan space, Sasha anticipated the glum atmosphere that usually pervaded the space. But she noticed something different today: a frisson or a buzz. Dan walked across from the office cooler, attempting to suppress what could only be good news judging by his taught facial expression – you could read Dan like a book.
>
> I've got good news – we've won the Broadmill contract – all that work on the bid was worthwhile and all thanks to your team, Sasha.
>
> Just in that moment she could feel her jaw drop, her eyebrows rise and her heart miss a beat. Sasha felt a groundswell of surprise and the release of two months of pent-up anxiety.

In this narrative, we can identify a whole range of feelings played out in the office environment. But whilst it is difficult to make sense of this melee of good and bad evaluative feelings, it is possible to make some incisive observations. This scenario illustrates three important subjective feelings that we experience in everyday life: emotion, mood and temperament. Each of them is intimately linked and share common characteristics. Emotions, moods and temperaments differ in several important respects (Gray and Watson 2003), as also illustrated in Figure 2.3:

- duration;
- frequency;
- comprehensiveness;
- intensity;
- pattern of activation.

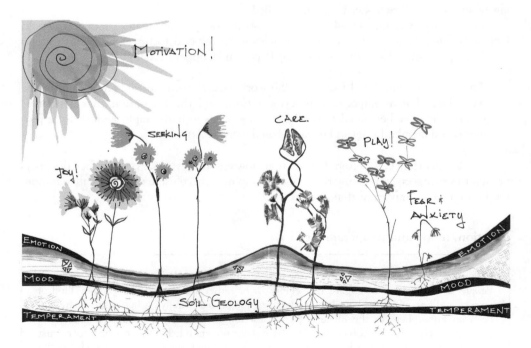

Figure 2.3 Emotion, mood and temperament.

Duration: Emotions refer to discrete sensations that are brought about by singular events. They happen in an instant of time. If we look at the scenario involving Sasha, we can envisage the emotion (including the facial expression and physiological change) that accompanied the good news. Mood, in contrast, is a state that occurs over a period of time (hours or even days). We can see in the scenario that a general mood of anxiety and apprehension had affected Sasha. We often talk about mood being infectious – and in this instance the whole workforce seems to have been affected by a mood of anxiety. Lastly, we can see the effect of temperament. Sasha, by her own admission, was predisposed to see the situation in a negative way. Her temperament represented part of her character: it was a permanent trait, or at least something that changed only very slowly over time. We think of temperament as being partly determined by our genes. Any change in temperament is something that happens over years rather than days. The duration of emotion is fleeting; the duration of mood is more prolonged; the duration of temperament is often lifelong.

Frequency: Whilst several different emotions might be experienced during the course of the day, mood changes less frequently. Often mood is associated with diurnal patterns associated with levels of hunger or tiredness.

Comprehensiveness: Whilst emotions might refer to a number of distinct feelings such as fear, anxiety and anger that occur at the same time, mood constitutes a 'summary' of all transient states that an individual feels at a given point in time. Generally, we attempt to summarise our mood as either good or bad. Mood represents our 'litmus test'.

Intensity: Emotions are experiences that typically are more intense than moods. For example, an emotion such as joy is typically attenuated into a general mood of cheerfulness and contentment. In the office environment, mood is a much more commonly experienced feeling, representing a 'sub-threshold' counterpart to an emotion. So, for example, people can feel annoyed and irritated (the sub-threshold version of anger); people can feel nervous and tense (a mild form of fear); or they can feel cheerful and pleasant (the attenuated version of joy).

Pattern of activation: Emotions are generally activated by specific stimuli (either endogenous or exogenous). This might include personal thoughts and reflections; an interaction with a co-worker; or perhaps a response to a feature in the built environment. In direct contrast, moods do not require any particular stimulus to be present. Instead they represent a summary of affective states indicating how we are feeling generally.

Equipped with a clear understanding of emotions, how does that affect our approach to office design, facilities management and customer engagement? What is clear is that the office environment can be used to stimulate emotions. It can also be used to influence moods, not through direct activation but by influencing the overall affective states of people, whether they be working individually or as teams. Finally, temperament is unlikely to be changed by the workspace we create. Each of us brings with us a distinct character largely determined by our genetic make-up. The challenge is for the designer to enable diverse personalities to thrive and engage in their preferred and distinct work style. Rather than 'influencing', the challenge is 'accommodating' or even 'nurturing'.

Emotion and the environment

What are the emotional qualities that we attach to our physical surroundings? Many words might come to mind like exciting, joyful, boring or pleasant. It seems that there are too many words to make sense of. An early study by Russell and Pratt (1980) attempted to

18 *Understanding emotions in the workplace*

create a simple framework to capture the many diverse affective responses to place. They found that other researchers had limited their investigations to only a few affective descriptors (often only one at a time). They wanted to find out whether a simple model could capture the abundant emotional responses they encountered when they asked people about their feelings about their surroundings. They encountered 105 common adjectives used to describe environments. They used the word 'affect' to refer to any emotion that was expressed in language, and the 'affective quality' of a place as the emotion-inducing quality that people verbally attributed to that space. Surprisingly, they found that just two factors were sufficient to capture all of the 105 adjectives (using factor analysis). These two factors were capable of characterising the spread of emotions in a two-dimensional space. The two factors were:

- Dimension 1: ranging from unpleasant to pleasant.
- Dimension 2: ranging from sleepy to arousing.

Figure 2.4 shows the two principal components and 21 clusters of adjectives identified by Russell and Pratt (1980).

When the different emotions were plotted on this 360° graph, they found that rather than clustering about the axes, the descriptors were distributed meaningfully throughout the space. Thus, for example, the word 'exciting' cannot be described as either arousing or pleasant alone. It constitutes some distinct combination of pleasantness and arousal.

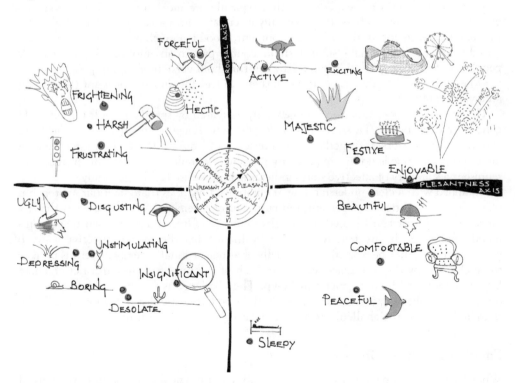

Figure 2.4 The affective qualities of space.
Source: Russell and Pratt 1980.

The 360° plot also indicates bipolar relationships (antonyms) such as exciting versus gloomy. One of the outputs from the analysis was a simple set of scales that could be reliably used to establish the affective quality of places. The approach used by Russell and Pratt (1980) has been widely adapted and is extensively used to capture people's feelings about space.

More recent research has proposed the replacement of the two-dimensional model of arousal and pleasure by a three-dimensional model. By introducing the concept of 'dominance' as a third dimension it is suggested that a more complete representation of human emotion is achieved (Bakker et al. 2014). Dominance refers to people's sense of control and the extent to which they feel their behaviour is restricted. ¶

Summary

This chapter examined the nature of emotions and the related concepts of mood and temperament. It indicated that our experience of emotions is a multistage process. Negative emotions have been the principal concern for designers of workspaces, seeking to alleviate the stress that abounds in modern work environments. Positive emotions have been much neglected, as has their potential to create 'flourishing' individuals. The next chapter explores whether the workspace can provide a motivating factor in the evolving world of work.

References

Bakker, I., van der Voordt, T., Vink, P. and de Boon, J., 2014. Pleasure, arousal, dominance: Mehrabian and Russell revisited. *Current Psychology*, 33 (3), 405–421.
Carlson, J.G. and Hatfield, E., 1992. *Psychology of Emotion*. San Diego, CA: Harcourt Brace Jovanovich.
Ellsworth, P.C. and Smith, C.A., 1988. From appraisal to emotion: Differences among unpleasant feelings. *Motivation and Emotion*, 12 (3), 271–302.
Estrada, C.A., Isen, A.M. and Young, M.J., 1997. Positive affect facilitates integration of information and decreases anchoring in reasoning among physicians. *Organizational Behavior and Human Decision Processes*, 72 (1), 117–135.
Fredrickson, B.L., 2001. The role of positive emotions in positive psychology: The broaden-and-build theory of positive emotions. *American Psychologist*, 56 (3), 218–226.
Fredrickson, B.L., 2003. The value of positive emotions: The emerging science of positive psychology is coming to understand why it's good to feel good. *American Scientist*, 91 (4), 330–335.
Gray, E.K. and Watson, D., 2003. Emotion, mood and temperament. In: *Emotions at Work: Theory, Research and Applications for Management*. Chichester: John Wiley & Sons.
Montag, C. and Panksepp, J., 2017. Primary emotional systems and personality: An evolutionary perspective. *Frontiers in Psychology*, 8, 464.
Payne, R.L. and Cooper, C.L., 2003. *Emotions at Work: Theory, Research and Applications for Management*. Chichester: John Wiley & Sons.
Russell, J.A. and Pratt, G., 1980. A description of the affective quality attributed to environments. *Journal of Personality and Social Psychology*, 38 (2), 311–322.
Seligman, M. and Csikszentmihalyi, M., 2000. Positive psychology: An introduction. *The American Psychologist*, 55 (1), 5–14.
Simon, H.A., 1967. Motivational and emotional controls of cognition. *Psychological Review*, 74 (1), 29–39.
Stanley, R.O. and Burrows, G.D., 2003. Varieties and functions of human emotion. In: *Emotions at Work: Theory, Research and Applications for Management*. Chichester: John Wiley & Sons, 3–19.

The New Era of Positive Psychology, 2012. Martin Seligman, TED Talk [online]. Available from: www.ted.com/talks/martin_seligman_on_the_state_of_psychology?language=en61766 [accessed 29 March 2019].

Zadra, J.R. and Clore, G.L., 2011. Emotion and perception: The role of affective information. *Wiley Interdisciplinary Reviews: Cognitive Science*, 2 (6), 676–685.

3 Can a workspace motivate us?

In the previous chapter we looked at what is meant by emotions. We found that there are three related but distinct phenomena, all three of which determine how we feel about ourselves and how we interact with the environment – temperament, mood and emotion. These three concepts are distinguishable in terms of how long they last and how intensely they are felt. Writers often talk about 'trait versus state', with temperament being a largely genetic trait associated with character, whilst mood and emotion reflect our current emotional state (affect). Can a workspace respond to our mood, our emotions or even our temperament – rather like a trusted friend? If not, can a workspace at least allow us to express our own character and changing feelings? Now that we have developed a wider understanding of emotions, this chapter examines the question, 'What makes us tick?' More specifically, we think about what motivates us at work and whether the office environment can play its part.

Form follows function

If we are talking about the 'modern office' it's useful to begin at the beginning. Perhaps the most enduring maxim encountered in architecture today is 'form follows function'. This guiding thought is a modern-day interpretation of 'form ever follows function' – an idea espoused by the architect Louis Sullivan as far back as 1896. It was later adopted by his assistant Frank Lloyd Wright, enabling him to shake off the conventions and rules that persisted around building design whereby 'form followed precedent'. In Sullivan's treatise entitled *The Tall Office Building Artistically Considered* he asked:

> ...how shall we proclaim from the dizzy height of this strange, weird, modern housetop the peaceful evangel of sentiment, of beauty, the cult of a higher life?
> (Sullivan 1896, p. 403)

He put forward the question:

> ...how shall we impart to this sterile pile, this crude, harsh, brutal agglomeration, this dark, staring of eternal strife, the graciousness of those higher forms of sensibility and culture that rest on the lower and fiercer passions?
> (Sullivan 1896, p. 403)

Sullivan appeared to be concerned that the 'modern office building' of 1896 was becoming "the joint product of the speculator, the engineer, the builder". He wanted to reassert the role of the architect, who should be able to cast aside "books, rules, precedents, or any such educational impediments to a spontaneous and 'sensible' result".

If we reflect on the key characteristics of a building espoused by the Roman engineer Vitruvius, a structure should exhibit three characteristics: *firmitas, utilitas, venustas* – that is, it must be solid, useful and beautiful. The idea of 'form follows function' clearly places *utilitas* (or utility) centre stage. Sullivan's (1896, p. 407) conviction was: "All things in nature have a shape, that is to say, a form, an outward semblance, that tells us what they are, that distinguishes them from ourselves and from each other." He didn't like the tendency of trained architects when faced with the problem of designing a skyscraper of introducing a variety of "separate, distinct, and unrelated buildings piled one upon the other until the top of the pile is reached" (1896, p. 407). He resented 'ornamentation' based on design catalogues or patterns and felt that beautiful design was an inevitable product of functional design.

Following his reasoning that every problem contains and suggests its own solution, he, like his contemporaries involved in tall building design, arrived at a blueprint based on the function of each storey. In fact, it was seen as one awe-inspiring machine. The first floor below ground level contained the plant room necessary for power, heating and lighting. The second floor provided the street level retail outlets including stores, banks and eateries. This floor needed to possess sufficient light, space and access to enable these functions. The floor immediately above the street level, being accessible by stairways, made use of large subdivisions to allow an expensive natural lighting through external openings. It was the floors above the second floor that presented a dilemma for many designers. The thought of many floors piled on top of one another to accommodate office space did not sit well with the design instinct for variety. He prescribed an office as similar to a "cell in a honeycomb, merely a compartment and nothing else" (Sullivan 1896, p. 404). Completing this 'circulatory system', he envisaged a top level that was related to the "life and usefulness of the structure" serving a purely physiological purpose, providing the "grand turn" in the circulatory system filled with tanks, pipes, valves and other mechanical equipment.

Clearly, we can see in this blueprint for tall buildings Sullivan's need to reinsert an emotional or design-led perspective. A radically new type of building was shifting the power base away from the architect to the structural engineer (steel frame construction) and the building services engineer (elevator and air-conditioning). Where was the emotion in all of this? His conviction was that "... we must now heed the imperative voice of emotion" (1896, p. 403). He felt that architects had to embrace change and be an integral part of the solution, rather than be confined to the periphery of old, established design principles. It seems that the profession had to be brought kicking and screaming into the twentieth century or else risk being consumed by the emerging technological professions. 'Form follows function' provided a way to legitimise the awe-inspiring machine aesthetic.

Form follows emotion

'Form follows function' still remains a universal design principle. But it seems that a new rule is replacing it as the dominant paradigm (see Figure 3.1). In 1981, a strategic design company called Frog collaborated with Apple to produce the highly original product – the Macintosh SE computer. Its company founder, Hartmut Esslinger, used a refreshingly new principle: 'form follows emotion'. He believed that "no matter how elegant and functional a design, it will not win a place in our lives unless it can appeal at a deeper level, to our

Figure 3.1 Form follows emotion.

emotions" (Sweet 1999, p. 9). Having previously spent time studying engineering, Esslinger became disillusioned with the students' approach. "They were all obsessed with clean lines, the machine aesthetic; they wanted to strip objects back to their basic function and make them emotion-free. It was so dull, I hated it" (Sweet 1999, p. 9). Having founded the company Frog, he recognised that the emotional element was present in consumer products in several different ways:

> It may appeal to our desire for enhanced nostalgia, as in the design for Dual electronics, or it might be a tactile ergonomic experience, as deployed in a computer game joystick that is sculpted to fit snugly in the hand. Or it could be in reinventing the familiar: the massive selling AT&T answering machine was flipped on its side with the dual emotional payoffs of showing consumer individuality and helping to keep a desk tidy.
>
> (Sweet 1999, p. 9)

The workspace environment is not a product. It is a relationship involving people and space. But the same adage 'form follows emotion' could become the new guiding principle. However, if we are to adopt this principle, we need to understand the full breadth of emotions that make us function at work.

Evolution of the office

Moving forward from the era of Sullivan to the office of the 1930s, and you see ideas about building intelligence starting to emerge. Le Corbusier commented:

> These skyscrapers will contain the city's brains, the brains of the whole nation. They stand for all the careful working-out and organisation on which the general activity is based.
>
> (Quoted by Banham 1980, p. 254)

The phrase 'the enormous file' was coined by Mills and Jacoby (1951) in a publication entitled *White Collar: The American Middle Classes*. The office came to represent a large, centralised bureaucracy that had to deal with the overwhelming surge in paper records. Division of labour and the appearance of dedicated office equipment was soon in evidence with the encroachment of technologies like comptometers, dictaphones and addressographs.

The continued survival of 'the office' defies many futurologists. The *Architect's Journal* in 1973 (cited by Delgado 1979, p. 11) noted:

> The concept of the office can be seen as one of the most consistent threads in any culture, for systems of government and manufacture may change beyond recognition, but in any organisation of human beings which extends beyond the smallest group, the word office, and the idea it represents, emerge as stable components of language.

Several discrete stages in the evolution of the office are distinguishable by the prevailing 'unit of currency':

- 'The library' (Renaissance) – the record is paramount.
- 'The coffee house' (seventeenth century) – the transaction is paramount.
- 'The enormous file' (early twentieth century) – the document is paramount.
- 'The intelligent space' (late twentieth century) – the screen is paramount.
- 'The emotionally intelligent space' (twenty-first century) – the person in front of you is paramount.

By the 1970s and '80s, startling changes start to emerge in the office:

> An electronic revolution of staggering magnitude is sweeping the office. The integrated electronic office is becoming a reality made possible through the merging of telecommunications and computer technologies. Filing cabinets, office memos, and written reports are being replaced by a network of computer terminals with a huge capacity for generating, storing, and processing data.
>
> (Lowe 1984, p. 137)

The changes associated with the electronic revolution did indeed transform the office landscape. But these changes pale into insignificance compared to those that are taking shape in the office of the twenty-first century. The advent of the Internet allowed the 'enormous file' to extend its reach well beyond the boundaries of the office building.

Even within the office building, people could hook up to the organisational brain wherever they pleased. But a more important change is now underway. It is no longer just about becoming more efficient through new technology. Instead of just being a cog in a huge machine, office workers have joined the ranks of 'creatives' – people who seek originality and expressiveness in their work, and who have the opportunity to develop their imagination. Even people with seemingly routine jobs now see themselves as creatives. Software developers, knowledge workers and project managers are just a few of the professionals who are liberated from routine process-driven work – who no longer 'feed' the machine – they design the machine. We are only just beginning to understand how the workspace affects this new type of employee. What makes them tick? What types of lever can stimulate creative work? What types of environmental distraction inhibits that elusive condition of creativity?

What makes people tick?

What makes people tick? Understanding emotions goes some way to explaining how people engage with their environment but it doesn't tell the whole story. After all, people are not passive receivers of emotional experience. There is a life-force or energy that activates people and directs their efforts. This goal-driven energy is what we call motivation. Our quest to understand the emotionally intelligent workspace would not be complete without understanding motivation. Motivation and emotion have a cause and effect relationship. We are driven by internal motives and at the same time external stimuli – to pursue particular rewards. We act on these motives in the hope of experiencing positive emotions or happiness. Whilst an emotion describes our state, our motivation describes our desire to change our state. Both concepts are related, and even the words originate from a common Latin root word *movere* (to move).

Franken (2006) defines motivation as the "arousal, direction and persistence of the person's behaviour". Motivation is the reason for people's actions, willingness and goals. One form of motivation known as intrinsic motivation describes the self-desire to seek out new things and new challenges, to analyse one's capacity, to observe and to gain knowledge.

We often think of both motivation and emotion as warm and personal experiences in contrast to cognitive functions which are perceived as cold and impersonal. Furthermore, they both involve a relationship between an individual and their environment.

Maslow's hierarchy of needs

An inevitable starting point on the subject of motivation is the hierarchy of needs proposed by Abraham Maslow (1943). In his paper A *theory of human motivation* he set out a framework to describe the needs or motivations that we all have. He focused his attention on 'exemplary' people that appeared to be thriving and fulfilling their full human potential: people that he described as positively healthy, highly evolved and matured, self-actualising. He criticised the prevailing obsession amongst psychologists to look at people who were mentally ill:

> It is as if Freud supplied us the sick half of psychology and we must now fill it out with the healthy half.
>
> (Quoted by Lomas 2016)

26 *Can a workspace motivate us?*

Central to Maslow's theory is the existence of a universal set of five distinct motives:

- physiology;
- safety;
- affection;
- esteem;
- self-actualisation.

The key feature of the theory is that motives can be organised using a hierarchy. In this hierarchical arrangement some motives take precedence over others. For example, if a person is starving, their physiological need to obtain food will override any other needs (motives). All of their resources and attention will be focused on satisfying this need. Once the person accomplishes the goal of satisfying their hunger, that person can then focus on 'higher-order' motives such as affection or esteem. At the top of this hierarchy is self-actualisation, which is concerned with fulfilling one's creative potential. The concept of 'cognitive priority' is often represented in management textbooks using a pyramid. Despite the fact that Maslow never used a pyramid to explain his model, researchers have suggested that "the powerful visual image of a pyramid of needs has been one of the most cognitively contagious ideas in the behavioural sciences" (Kenrick *et al.* 2010).

In Figure 3.2, an alternative to the pyramid metaphor is used: an elevator rising up through a building. Cognitive priority requires that you have to go in through the entrance before ascending to the different storeys of the building. Using this metaphor, we relate the hierarchy of needs and its fulfilment to building design (see Finch 1992).

Maslow's hierarchy captures in one diagram the struggle to create an emotionally responsive building. At the bottom of the hierarchy are the basic needs that we require in order to survive (physiological) and function (functional). At the top of the needs hierarchy is self-actualisation – a term used to describe a state of creative fulfilment. In between, we see that there is increasing attention to the psychological well-being of the individual and their social participation. Maslow envisaged that everyone was engaged in a struggle to attain the top level of fulfilment. As to whether a workspace could help in satisfying these higher needs, Maslow was quite dismissive. Whilst he recognised that the physical environment serves basic human needs for shelter and security (physiological), he believed that ultimately the environment only serves to impede psychological growth and autonomy and therefore has to be 'transcended'. Indeed, it seems that modern-day work environments struggle to go beyond simply satisfying basic needs.

Does Maslow's hierarchy of needs have any relevance to modern-day workplaces? In the following sections we look at each of the needs in turn. We examine whether the basic needs stand up to the scrutiny of modern-day research. We also illustrate how the modern-day workspace can satisfy these needs.

Physiological needs

This is a showstopper! Without the ability to satisfy basic survival needs, the human condition cannot be sustained. This played a fundamental part in our ancestral past as we responded to feelings such as hunger or thirst. These are examples of homeostatic needs which arise when our physical condition goes 'out of balance'. We have all inherited an adaptive response that allows us to withstand fluctuations in the external environment through a series of feedback loops. In the modern workplace, we design environmental

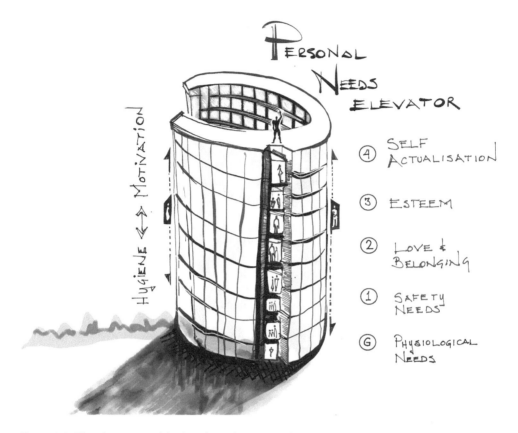

Figure 3.2 The elevator model of Maslow's human needs.

systems that reduce the need for adaptive behaviour (for example, we don't expect to modify our levels of clothing according to the fluctuation of temperature in the office). Basic human needs are not confined to hunger and thirst. In the office environment indoor air quality, thermal comfort and adequate lighting are taken for granted. Only once these basic physiological needs are met can higher-order needs be entertained.

Safety needs

Whether we design or manage buildings, safety concerns are always uppermost in our minds. But safety needs are as much about 'perceived' safety as they are about actual safety. It is the perception of safety within the minds of employees that will determine how people respond to the workspace. An unpredictable and unstable setting will inevitably drain people's mental resources and inhibit creative thought.

If we think about how our brains are hardwired, we are continually making a trade-off between 'peril' and 'prospect'. Every new encounter presents a possible opportunity or threat – and is associated with a desire to flee or confront a situation. In the work environment, our attentional systems are continually drawn to possible threats around us. We have learned to rapidly associate likely causes of threat using a complex filtering system.

For example, we are able to interpret other people's angry facial expressions. Our feelings of security may be threatened by noise, unfamiliarity and encroachment onto territory. What happens in an open plan work environment? We have learned to accept changing faces around us without feeling threatened. This is very different from our primordial reaction to unfamiliar surroundings that compromise our feelings of safety. We have learned to suppress this primitive response, but this does not occur without some drain on our physical and cognitive resources.

Taken together, physiological and safety needs represent the bedrock of any workable office. They satisfy our 'survival' needs. But people are not content with the satisfaction of these two lower-order needs. After all, people don't get out of bed in the morning driven by the desire to work in a safe place. This argument was put forward by Herzberg (1966) who described these two lower-order factors as 'hygiene' factors. What really makes people tick – so the argument goes – is the desire to fulfil higher level psychological aspirations such as achievement, recognition, responsibility and advancement. He described these factors as 'motivating' factors in his Two-Factor Theory. In essence, he stated that there are some factors in the workplace that can cause dissatisfaction and an entirely different set of factors that cause job satisfaction. This is an important observation because it reminds us that no matter how much we try to overcome dissatisfaction in the workplace (i.e. by meeting physiological or security needs), none of these hygiene factors will lead to job satisfaction. Poignantly, the physical environment was included amongst the dissatisfiers – alongside company policy and administrative practices, supervision, interpersonal relationships and salary. But if you are involved in workplace design this conclusion is unedifying. It suggests that we are only ever capable of tackling homeostatic needs (such as ensuring that the temperature is right or the noise levels are under control). Satisfaction surveys often only look at 'hygiene' factors (i.e. they are dissatisfaction surveys). Can the physical workspace be included amongst the motivating factors rather than being relegated to simply a hygiene factor? Can Herzberg and Maslow be proved wrong?

Affiliation and belongingness needs

Now we move on to Maslow's higher-order needs that are not essential to survival but which inject meaning into our everyday life. As humans we have evolved as social beings – reflecting the fact that our ancestors lived in groups as hunter gatherers. Today the desire for love, affection and belonging are an essential part of our existence. So sensitive are we to these needs that any form of social rejection creates a response in our neural circuits that resembles that of pain. Social interaction also leads to the release of the brain chemical oxytocin which in turn leads to feelings of warmth and well-being. This desire to belong has an evolutionary purpose. It promotes the sharing of resources, the exchange of knowledge and, in everyday life, the sharing of parenting chores. Modern-day studies have identified functional and neurological differences between three different types of affiliation: romantic, family membership and friends. There is a case for identifying separate needs for each of these three types of affiliation on Maslow's hierarchy.

How does our understanding of belongingness colour our approach to workplace design? The concept of hot-desking has often been the subject of scrutiny when we think of people as social animals. On the one hand, hot-desking can be seen as a way of breaking open an organisation and creating the opportunity for new unplanned interactions. Critics of the system would say that it fractures established social networks and only creates a weak, fragile substitute. Many other interventions are at the disposal of designers and managers: some of these can enhance group affiliation whilst others can unwittingly corrode social connections (Sennett and Kovalainen 2000).

Status and esteem needs

Surely status has no part to play in the modern-day office? After all, providing a larger office seems to be a very expensive way of rewarding people, as well as a very divisive one. But perhaps we should consider the adaptive role of esteem needs in our evolutionary past. Esteem describes two distinct forms that play out in the office:

- self-esteem (which reflects a personal desire for strength, achievement and mastery);
- esteem of others (reputation, status and dominance).

Acquisition of a skill makes an individual more indispensable to a group. We strive for mastery at our work, knowing that this will confer status. We have something that other people value. In an information-based economy people know that they are able to learn from an expert. They express deference towards an individual with special skills. This deference reduces conflict with a dominant person.

In order to establish status, we are inclined to seek novel information or explore alternative solutions to the problem. This process of mastery enables the person to receive favours from other people because of their special skills.

Schadenfreude

Ellen has walked right into it. She's a 'newbie' and has set herself up this morning at the hot-desk that is definitely not 'up for grabs'. It's located next to the scenic window and the desktop provides the only Reuters feed on the office floor. More importantly, it's the desk that is routinely used by one of the senior partners, Maggie Wainwright. There are unspoken rules here! In the past it was so much easier – you couldn't ignore the nameplate on the door. In the open plan office rules still exist, but they are tacit. It's always entertaining to watch the newbie blunder through this forest of unspoken rules: knowing that they will have to learn the rules of territory the hard way.

Self-actualisation

Let's imagine that we've managed to travel up in our elevator to the top floor. At this point in our ascendancy we have satisfied all our basic human needs – except one! Maslow suggested that a new discontent and restlessness emerges even when all the preceding needs are satisfied. Individuals still have a need to "do what they are fitted for. A musician must make music, an artist must paint, a poet must write" (Maslow 1943, p. 372). He described this universal need as 'self-actualisation', positioned at the very top of the needs hierarchy. To become self-actualised is "to become everything that one is capable of becoming".

> One thing I'm really interested in … and this is something that comes up in our team-day tomorrow … is this whole idea of intrinsic motivation. This job feels different. There's more of me in this. When we are able to tap into people's intrinsic motivation, that's when the best of who they are comes out. Not many workplaces do that. If you can tap into someone's meaning there's going to be a whole lot more comes out of that.
>
> Consultant describing their plans for a start-up space

Criticisms of Maslow's hierarchy of needs

Has Maslow's hierarchy of needs stood the test of time? As stated by Kenrick *et al.* (2010), "many behavioural scientists view Maslow's pyramid as a quaint visual artefact without much contemporary theoretical importance. We suggest, on the contrary, that the idea can take on a new significance when combined with later theoretical developments." As Maslow pointed out, the theory represents a "framework for future research and must stand or fall, not so much on facts available or evidence presented, as upon researches to be done, researches suggested perhaps, by the questions raised in this paper" (Maslow 1943, p. 372).

ERG theory

A more robust and simplified version of Maslow's hierarchy of needs was put forward by Alderfer (1969), a theory that makes use of a three-tier model of human needs. Known as the ERG theory, it identifies 'existence', 'relatedness' and 'growth' as the three key stages of human fulfilment.

1. Existence needs: refers to all material and physiological requirements (e.g. food, water, thermal comfort, safety (this embraces Maslow's first two levels).
2. Relatedness needs: encompasses the need for relationships with co-workers and employees. It includes feeling recognised and secure as part of a group.
3. Growth needs: includes our desire for internal esteem and self-actualisation. It embraces the fourth and fifth levels of Maslow's hierarchy. It describes the human compulsion to be creative or productive.

We are accustomed to a work environment that satisfies our 'existence needs'. After all, who can do a day's work feeling too hot, feeling hungry or even feeling unsafe? Undoubtedly buildings have become much more adept at meeting our basic existence needs. The advent of comfort conditioning for temperature control, passive and active noise control, lighting control, and more-resilient fire and safety systems are all examples of homeostatic systems that reduce the stresses imposed on office workers. More complex intelligent buildings have allowed people to concentrate on higher needs.

Organisations are beginning to recognise the importance of 'relatedness needs' and increasingly understand the value of architecture as a way of 'engineering' communication in the workplace. For example, Zappos, the online clothes retailer (Waber *et al.* 2014, p. 11), have introduced a new metric called 'collisionable hours' to reflect changes in interaction levels. However, our understanding of interaction remains rudimentary – as do our efforts to coerce people to engage within the workspace. We might have a detailed knowledge of behaviour and interaction (often enabled using tracking technologies) but it still doesn't allow us to get inside the minds of people involved.

Moving one level up, we arrive at 'growth needs'. This typically involves periods of high concentration on a task. When it comes to office design, satisfying this level is often in direct conflict with satisfying relatedness needs. Whilst buzz and high-energy might help with binding social fabric, it does little for the person trying to undertake 'deep work' (Newport 2016).

Summary

Everyone wants a job that is fulfilling – that allows us to be the very best that we can be. We like to surround ourselves with the team that will enable us to realise our potential. We also want to belong in a space that makes that possible – or at least does not inhibit our growth. Looking again at our hierarchy of needs, the office of today can go well beyond existence needs, enabling us to function as part of a social group (relatedness needs) and to flourish as individuals (growth needs).

Today, information is not the only common thread. No longer does the office have the monopoly on information – it can be obtained from almost any location. It is the transformative possibilities that arise from having people co-located that sets the office apart from other settings. The success of the office rests on its ability to leverage human input. In the next chapter we look at the technological perspective of building intelligence. By tracking its evolution, we reflect on the precarious relationship between technology and emotion.

References

Alderfer, C.P., 1969. An empirical test of a new theory of human needs. *Organizational Behavior and Human Performance*, 4 (2), 142–175.

Banham, R., 1980. *Theory and Design in the First Machine Age*. Cambridge, MA: MIT Press.

Delgado, A., 1979. *The Enormous File: Social History of the Office*. London: John Murray Publishers.

Finch, E., 1992. Facilities management at the crossroads. *Property Management*, 10 (3), 196–205.

Franken, R., 2006. *Human Motivation*. Sixth Edition. Florence, KY: Thomson/Wadsworth.

Herzberg, F.I., 1966. *Work and the Nature of Man*. Oxford: World.

Kenrick, D.T., Griskevicius, V., Neuberg, S.L. and Schaller, M., 2010. Renovating the pyramid of needs: Contemporary extensions built upon ancient foundations. *Perspectives on Psychological Science*, 5 (3), 292–314.

Lomas, T., 2016. Why so serious? The untapped value of positive psychology [online]. *The Conversation*. Available from: http://theconversation.com/why-so-serious-the-untapped-value-of-positive-psychology-61766 [accessed 29 March 2019].

Lowe, G.S., 1984. "The enormous file": The evolution of the modern office in early twentieth-century Canada. *Archivaria*, 19, 137–151.

Maslow, A.H., 1943. A theory of human motivation. *Psychological Review*, 50 (4), 370–396.

Mills, C.W. and Jacoby, R., 1951. *White Collar: The American Middle Classes*. New York: Oxford University Press.

Newport, C., 2016. *Deep Work: Rules for Focused Success in a Distracted World*. London: Hachette UK.

Sennett, R. and Kovalainen, A., 2000. Book reviews: "The Corrosion of Character: The Personal Consequences of Work in the New Capitalism". *Acta Sociologica*, 43 (2), 175–177.

Sullivan, L.H., 1896. *The Tall Office Building Artistically Considered*. Los Angeles, CA: Getty Research Institute.

Sweet, F., 1999. *Frog: Form Follows Emotion*. London: Thames and Hudson.

Waber, B., Magnolfi, J. and Lindsay, G., 2014. Workspaces that move people. *Harvard Business Review*, 92 (10), 68.

4 The many faces of the intelligent building

The advent of the so-called 'intelligent building' has captured the imagination of technologists and science-fiction writers alike. The idea of a building possessing a brain connected throughout by innumerable sensors and actuators created a vision of unbounded possibilities. For some this aroused concerns about 'big brother' and the surveillance society. In practice, much of the research behind intelligent buildings in the latter part of the twentieth century was driven by a need to control the ad hoc arrival of new technologies in the workplace. Without common standards, it soon became evident that, rather than supporting the workplace, new technology hampered productivity. Devices were unable to talk to one another. Whilst they appeared to possess a common language of '1's and '0's the reality was very different. A Tower of Babel was beginning to emerge, with different manufacturers adopting their own bespoke digital language. Combine this with the burgeoning range of technologies and you were left with a behemoth – a multiheaded monster. Cables wormed their way chaotically throughout buildings and threatened to consume all the available space between floor and ceiling. Cooling requirements soared with the intensification of space and heat produced by office equipment. The intelligent building promised a much-needed response to this ad hoc intrusion. Simultaneously, we saw the emergence of facilities management, a profession that promised to bring order to the chaos.

This chapter examines the successes and limitations of the intelligent building. Whilst new technology has enabled convergence; miniaturisation; mobilisation and modularisation, it seems to be lacking a 'soul'. The chapter considers whether an intelligent building is directly at odds with an 'emotionally intelligent building'. In other words, does new technology stand in the way of meaningful person-to-person and person-to-environment interaction?

The death of permanence

It seems that we are continually reminded that change and impermanence are a fact of life. As observed by Alvin Toffler (1970, p. 55) in his far-reaching book, *Future Shock*:

> The shift towards transience is even manifest in architecture – precisely that part of the physical environment that in the past contributed most heavily to man's sense of permanence ... We raze landmarks. We tear down whole streets and cities and put new ones up at a mind-numbing rate.

The technology-laden intelligent building is the epitome of impermanence. It exhibits what Toffler (1970) referred to as the economics of transience, which is based on three characteristics.

1 Advances in technology lower the cost of manufacture much more than the costs of repair work.
2 It is possible to incorporate improvements to a system/component as technological advances accrue over time.
3 Uncertainty over the future and the inevitability of changing needs encourages designers to 'hedge their bets'. Rather than invest in fixed forms and functions, we build instead for short-term use.

Layers within layers

The advent of the impermanent building structure required a fundamental rethink about the whole design process. No longer was it about creating monolithic structures. Even the tall building required a re-examination in the face of irrepressible change. A particularly apposite model known as the 'layering' (shearing) model, developed by Duffy (1990), reflected this change in thinking brought on by new technology. The model identifies distinct temporal and spatial elements that go to make a building. It recognises that a building is not just a single lifecycle. Instead, it is made up of multiple elements each with their own lifecycle characteristics. These include:

- site – describes the land and footprint of the building (indefinite lifespan);
- shell – describes the structure and outer fabric of the building (lifespan of 40–200 years);
- systems – describes the mechanical and electrical systems required to maintain comfort, lighting, security and data requirements (lifespan of 5–20 years);
- scenery – refers to the flexible partitions and furniture systems (lifespan of 5–10 years);
- set – concerns all the day-to-day elements that are routinely changed and adapted to meet users' requirements.

Each of these physical systems are distinct in several important respects.

- They each have their own lifecycles.
- They each involve different replacement decision points.
- They vary in terms of their intractability (i.e. you may be stuck with some decisions for a long time).
- The financial implications of each 'layer' may be quite different. We are inclined to think of site and shell as being the largest cost items through the life of a building. But if you factor in the replacement cycles that occur for short-lived elements, the costs associated with systems and scenery can eclipse costs associated with the building structure.

The layered model of Duffy provided an indispensable reminder of *Future Shock*. The model is just as poignant today as when it was first conceived. With the advent of ever shorter leases, incessant resizing, relocation and changes in tastes, building economics has changed irrevocably. Obsolescence and impermanence are the inevitable consequences of this changing landscape.

34 *The many faces of the intelligent building*

The emergence of the intelligent building (IB)

If we are to investigate the 'emotionally intelligent building', we need first of all to consider its counterpart: the intelligent building (IB). Its evolution has not been without trials and tribulations. Even today there are many concerns that remain about the intelligent building as a concept. Issues of privacy and control are concerns that do not go away. However, there have been many advances that have made aspects of building intelligence an indispensable part of the modern work environment. If we look at the emergence of the intelligent building, we can see pivot points where the direction of motion has changed. The intelligent building has redefined itself several times over. Rather than present a single definition, it is perhaps more instructive to track its evolution (see also Figure 4.1):

- the high-tech building (1980–1990);
- the flexible building (1985–2000);
- the interoperable building (1995–2010);
- the green building (2000–present).

The high-tech building

Many people might describe it as the era of the 'wild West'. A bonanza of new technologies was emerging in the workplace. Alongside the desktop computer came a plethora of devices to support business communication. But each of these devices needed to be

Figure 4.1 The emergence of the intelligent building.

connected – if only for power. Soon the workplace began to resemble a swamp of hazardous cables and noisy heat-generating technology. Whilst these technologies promised advances in productivity, the disruption to human activity compromised progress. For the first time, organisations like Steelcase and Herman Miller recognised the need to address this problem: how could new technology that was appearing in the workplace live in harmony with office workers? Desking systems, furniture systems, partitioning systems and cabling systems formed part of an emerging array of 'scenery' designed to assimilate new technology in the open plan landscape. In tandem with this costly investment arose the need for a workplace professional that we know today as the facilities manager. They had the requisite skills to procure, install and manage modular systems that addressed technological and human needs.

At the same time as digital business emerged in the workplace, we also witnessed the advent of sophisticated but single function building automation systems. These single dedicated systems provided safety, access control, lighting; lifts and HVAC (heating, ventilation and air conditioning) capabilities. Each of these dedicated systems relied on feedback loops using sensors, controllers and actuators. A human operative could oversee operations and intervene where necessary. Many office users felt their ability to control their local environment had been compromised.

It was an era of trauma akin to the 'railway mania' experienced in 1840s' Britain or the Klondike goldrush of 1896 in the USA. In a frenzy to implement new building technologies, minimal attention was paid to standards. Tanenbaum (2002, p. 254) ironically observed:

> The good thing about standards is that there are so many to choose from.

Imagine, if you will, the absence of standards in a rail network – in particular, differences in rail gauge between competing lines. In the free-for-all that emerged amongst rail investors in Victorian Britain, there were many occasions where track width (train gauge) varied between operator. If you were unfortunate enough to be a passenger or if it was carrying your freight, a break of gauge could be very unwelcome. It caused delays, cost and inconvenience. Much the same problem occurred with electronic communications and building automation systems. Each dedicated system required its own cabling network that was required to run throughout the building. Every system used in the building relied on a different communication protocol. Far from being an intelligent building, the bespoke and proprietary systems locked-in clients to specific suppliers or manufacturers. During a building's operational phase, the anticipated efficiencies were not realised.

Cynics might suggest that this absence of communication standards served the interests of many suppliers and manufacturers. Building owners were often tied to one solution throughout a building's life without any opportunity to switch to another solution. In practice, the net effect was to have suppliers jostling for a larger slice of what was a very small pie. The early promise of an 'intelligent' building or smart building was undermined by ad hoc innovation. Whilst the individual innovations in building management systems were pioneering, the inability of systems to work together meant that the emergence of the intelligent building was thwarted:

> Over the past 20 years, many different buildings have been labelled as intelligent. However, the application of intelligence in buildings has yet to deliver its true potential.
> (Clements-Croome 2004)

Rapid organisational change also exposed the limitations of buildings in general. Not only did organisations change in size, the various business units within an organisation also expanded and contracted at unprecedented rates. By the mid-1980s the ownership of commercial property had begun to feel like a burden. After all, who wanted to be tied to a property portfolio that increasingly seemed out of touch with an organisation's space needs. An intelligent building now more than ever needed to be flexible enough to accommodate changing needs at the macro and the granular level.

The flexible building

By the early 1980s things were getting out of hand. Business demanded new technology in the workplace but buildings were unable to accommodate these demands. Whilst the technology was becoming increasingly sophisticated, existing building stock struggled to keep pace with requirements. Office buildings as young as 10 years of age were made obsolete by the blinding fury of technological change. Key factors affecting technological obsolescence included:

- the need for high volumes of uninterrupted telecommunications with the outside world;
- increased demand for power, including vertical and horizontal distribution across floors;
- the presence of environmentally demanding IT equipment that required a re-examination of floor capacity, humidity, cooling, acoustics, static control and vibration;
- the need to change the location of workstations, equipment and cabling.

Intelligent buildings were defined by their ability to accommodate new technology rather than the possession of technology in its own right. This sea-change in thinking was led by thinkers behind the Orbit report, a combined study involving North American and UK organisations.

ORBIT-2 (Becker 1988) provided a building rating process that allowed specifiers to assess just how 'future ready' a building was. Facilities managers were encouraged to "know your building's IQ". The model pivoted around a demand-side and supply-side understanding of an organisation. What did the organisation need? What did the organisation possess? By carefully addressing the changing needs of an organisation it was possible to devise a 'contingent' model that was based on a 'best fit'. The demand-side review demanded a 'future oriented' reflection of emerging issues affecting work styles. The low-change/routine organisations were contrasted with high-change/non-routine organisations whose IT requirements were markedly dissimilar.

If an intelligent building is the antithesis of a 'not-so-smart' building, the best-fit approach provided an indispensable method for distinguishing between the two. The building's design characteristics were as important as technology in realising an intelligent building. Floor span, floor-to-ceiling height, floor depth and risers all presented embedded design characteristics that would dictate whether a building was relegated to the top or bottom of the class. In an era of hardwired communications, the floor voids were becoming ever deeper and ceiling plenums ever larger. For the real estate industry, some buildings were becoming rapidly obsolete.

The interoperable building

The interoperable building provided the antidote to the uncooperative high-tech building. Integration became the buzzword. Rather than having dedicated building services operating in lonely out-backs within the building, a common protocol allowed conversations to occur in a mature, open and connected environment: no longer marooned islands of information, the intelligent building was able to harness intelligence from every device. This enabled more than just regulation of the building – it prompted the emergence of 'smart energy grids'. Interoperability meant that building automation and control systems (BACS) used a common language whose rules defined:

- data exchange – how fast and how often data was needed;
- alarms and events – how the events are logged and displayed;
- schedules – how HVAC plant and equipment operate as part of a schedule;
- trends – in the collection and processing of data and eventual display;
- network management – what type of diagnostic and maintenance functions are required and where they are located.

Interoperability had direct consequences for buildings and their capacity to accommodate IT. The convergence of building automation systems (BAS) and integrated communication systems (ICS) enabled the 'cable jungle' in the floor void to disappear. At the same time, the replacement of copper cabling by fibre optic cabling significantly reduced the floor volume occupied by cables. The advent of wireless Wi-Fi communication further reduced pressures on floor-to-ceiling height. Seemingly obsolete and unintelligent buildings could now become intelligent.

The green building

Intelligent buildings were initially conceived as 'technical fixes' to remedy all of our workplace comfort requirements. But this was often achieved at the expense of profligate energy consumption. So-called smart buildings were part of the problem rather than the solution. HVAC is a key part of an intelligent building that ensures thermal comfort and indoor air quality. The cooling demands of modern-day offices have an increasingly marked effect on the global demand for electricity. The energy they consume is likely to triple between now and 2050 in response to an increasingly warm planet. By 2050, the world's air conditioners are likely to consume the current electricity capacity of the USA, the European Union and Japan combined (Baraniuk 2018).

Many technological solutions may hold the key to our insatiable demands for office cooling. Examples include the development of 'nano-photonics' that make use of wafer thin, highly reflective material that radiate heat (at a wavelength that escapes the earth's atmosphere into space) even in direct sunlight. This allows water – cooled on average to a few degrees lower than the outside air temperature – to flow through pipes and be used to cool a building. Other currently available technologies such as inverters (that adjust the intensity of the cooling based on sensor readings) offer the promise of efficiency savings of 30–50 per cent.

By the 1990s, the definition of intelligent buildings was expanded to encompass users, building systems and the environment (Ghaffarianhoseini *et al.* 2016). It reflected a growing discontent with the technological definition of intelligent buildings. Instead, the focus turned to their effectiveness and efficiency, and their potential to respond to social and technological changes (Clements-Croome 2004). This realignment towards user interaction and social change was expressed by the engineering firm Arup as "one in which the building fabric, space, service and information systems can respond in an efficient manner to the initial and changing demands of the owner, the occupier and the environment" (cited by Brignall 2003).

The arrival of the 'sentient building'

The evolution of the intelligent building took an abrupt turn as we entered the age of the Internet. Rather than relying on the cognitive powers of the 'brain', intelligence is now a disparate set of technologies reporting to many different cores – both within and outside the building. This includes embedded objects within buildings forming part of the Internet of Things (IoT) and remote monitoring from distant locations. The building is no longer an enormous file or even an enormous electronic file. Accompanying this technological change is a redirection of purpose. For once, buildings have the capacity not only to understand themselves but also more importantly to understand the people that occupy them. We have now witnessed the advent of the 'self-aware' or 'sentient' building. Figure 4.2 illustrates the three evolutionary paths of buildings according to affect, behaviour and cognition. It highlights the relationship between the intelligent, sentient and emotionally intelligent workspace.

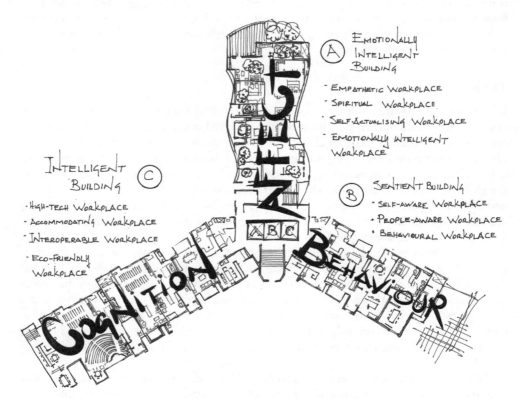

Figure 4.2 The ABC path of office evolution.

At the heart of a sentient building is 'ambient intelligence' – a term that has received increasing interest from smart-building devotees. The underlying aim of ambient intelligence is to "support human contact and accompany an individual's path through the complicated modern world" (Weber and Rabaey 2005, p. 1). The characteristics of ambient intelligence include:

- the technology is hidden in the background – being embedded in the surroundings;
- it is sensitive to the presence of people and objects – providing an appropriate response;
- it is able to provide assistance using smart 'augmented' technologies;
- it ensures data privacy and security whilst providing information to the user when it's required.

Ambient intelligence relies on small discreet devices that are low-cost and lightweight. They are often wireless systems that are able to communicate with one another. They have a level of redundancy so that the failure of one device does not cause failure in the whole system.

As a result of the sentient building, the science of environment-behaviour has been transformed. New technologies enable human behaviour to be tracked and analysed on an unprecedented scale. It's no longer about measuring the traditional real estate metrics of cost per square foot. Instead, the focus has turned to metrics that can measure how design can help or hinder organisational performance. The sentient building uses an ad hoc arrangement of sensors capable of reporting to a multiplicity of manufacturers and designers located off-site. Unlike the centralised intelligent building, the sentient building reaches out its tentacles to users. It uses disparate and improvised intelligence gathering from many different sources. As well as the traditional sensor technology, the use of activity trackers, smart phones and social networks allow performance data to be pooled at a level not seen before. Instead of reliance on time-consuming post-occupancy evaluations, real-time analysis using network analytics provide the measurement data required to prove whether a design works. Spaces become engineered spaces. No longer do facilities managers simply allocate space based on a prescribed space budget: data has to inform decision-making. Where do people go? Who do people meet up with? How long for? With what? All of this depends upon the agreement of building users, who may accept the risk of data misuse and intrusion if it leads to a better work environment.

On collision

At the centre of this environment-behaviour revolution is human interaction – more specifically, the merits of increased interactions or collisions. It seems that we are four times more likely to communicate regularly with someone sitting just 2 metres away than with someone 20 metres away. This was the finding of Thomas J. Allen in his 1977 book *Managing the Flow of Technology* (Allen 1984). We might believe, with the advent of electronic communication in the intervening time, that this correlation between distance and communication no longer holds. More recent research using sociometric badges confirms that both face-to-face and digital communication follows this rule of diminishing communication with distance (Waber *et al.* 2014). Those who shared a physical office were 20 per cent more likely to stay in touch digitally than those who worked elsewhere. Even more surprisingly, co-located co-workers emailed each other four times more frequently than colleagues in different locations.

People interactions have now attracted interest from many organisations. The practice of 'relational analytics' involves the science of human social networks (Leonardi and Contractor 2018). Employees leave behind a 'digital exhaust', a trail of digital information (emails, tracked physical movements, chats, phone calls) which can be routinely analysed, for example mapping the communications between two people during the course of a day. In addition to individual characteristics, the relationships that employees have with one another can supposedly explain workplace performance. Rather than looking at the transactions from within, the relational analytics approach seeks to map patterns or signatures "…just as neurologists can identify structural signatures in the brain's networks that predict bipolar disorder and schizophrenia" (Leonardi and Contractor 2018, p. 5). The six common signatures that they have identified are: (1) ideation; (2) influence; (3) efficiency; (4) innovation; (5) silos; and (6) vulnerability. The technology falls short of an emotional understanding of users, but it does enable a routine analysis of people–environment interactions – and in turn the identification of functional or dysfunctional space.

We know much more about office behaviour but almost nothing about why these behaviours exist. Sociometric and tracking tools undoubtedly provide basic insights into motivation and behaviour – humans as lab rats. These are driven by an instrumental view of office design that tells us that increased interactions are good for productivity. Not just positive interactions – interactions in general. But if we are to break open true human potential then we need to probe further.

Emotion detection

Emotion detection is one of the capabilities being explored using ambient intelligence. Health environments have attracted particular attention. A proof-of-concept project by a team in Spain (Fernández-Caballero *et al.* 2016) examined the use of emotion detection and regulation in smart health environments. Using a combination of face recognition and physiological measurements, they tested a smart environment (SE) to improve the quality-of-life for ageing adults at home. An electrodermal response sensor (EDRS) and a heart rate sensor were used to establish how unpleasant (valence) a stimulus is and how much excitement or calmness (arousal) it produced. By varying the colour/light and music in the environment, they experimented with 'emotional regulation'. Similar studies are likely to emerge in workplace environments, although concerns about ethics remain.

Summary

We have moved on from the high-tech; the flexible and the interoperable intelligent building. Now buildings have become enormous data gatherers and distributors. Instead of being the brain, the building has become the eyes, ears and feelers capable of self-awareness. This extends beyond simply monitoring air quality or lighting levels. In the distributed world we can track and interpret human behaviour. In isolation, individual tracking information provides a limited insight into building performance. But when viewed as part of 'mass data', compelling patterns emerge that are invaluable when used on an ad hoc basis by space consultants or increasingly on an ongoing basis as part of relational analytics.

But this evolutionary picture of smart buildings is not the end of the story. Human adaptation has been far outstripped by the rate of change imposed by new technology, so it could be said that new technology has neglected and perhaps abandoned the 'spirituality' that we have often taken for granted in buildings.

Are we looking for something more from our everyday work environment? Does building intelligence simply mean more technology? As observed by Clements-Croome (1997, p. 1):

> ...intelligence is becoming an overrated word which can be used to describe buildings, cameras or car cockpits. We are not even sure what human intelligence is, so how can we ascribe this description to products? It is possible that people will reject some of these forms of intelligent hardware in favour of using their own creative impulses.

He goes on to suggest that "technology must enhance the opportunity to explore not usurp human creativity" (1997, p. 1). In the next chapter we tackle the issue of emotional intelligence head-on. What might an emotionally intelligent workspace look like – with or without technology?

References

Allen, T.J., 1984. *Managing the Flow of Technology: Technology Transfer and the Dissemination of Technological Information Within the R&D Organization.* Cambridge, MA: MIT Press.

Baraniuk, C., 2018. How trying to stay cool could make the world even hotter. *BBC News*, 19 June.

Becker, F., 1988. The ORBIT 2.1 rating process. *Facilities*, 6 (3), 5–7.

Brignall, M., 2003. Set course: Intelligent buildings. *Guardian*, 3 May.

Clements-Croome, D., 1997. What do we mean by intelligent buildings? *Automation in Construction*, 6 (5), 395–400.

Clements-Croome, D., 2004. *Intelligent Buildings: Design, Management and Operation.* London: Thomas Telford.

Duffy, F., 1990. Measuring building performance. *Facilities*, 8 (5), 17–20.

Fernández-Caballero, A., Martínez-Rodrigo, A., Pastor, J.M., Castillo, J.C., Lozano-Monasor, E., López, M.T., Zangróniz, R., Latorre, J.M. and Fernández-Sotos, A., 2016. Smart environment architecture for emotion detection and regulation. *Journal of Biomedical Informatics*, 64, 55–73.

Ghaffarianhoseini, Amirhosein, Berardi, U., AlWaer, H., Chang, S., Halawa, E., Ghaffarianhoseini, Ali and Clements-Croome, D., 2016. What is an intelligent building? Analysis of recent interpretations from an international perspective. *Architectural Science Review*, 59 (5), 338–357.

Leonardi, P. and Contractor, N., 2018. Better people analytics. *Harvard Business Review*, 11 (1) (November–December), n.p.

Tanenbaum, A., 2002. *Computer Networks.* Second Edition. Upper Saddle River, NJ: Prentice-Hall.

Toffler, A., 1970. *Future Shock.* London: Bantam.

Waber, B., Magnolfi, J. and Lindsay, G., 2014. Workspaces that move people. *Harvard Business Review*, 92 (10) (October), 68.

Weber, W. and Rabaey, J., 2005. *Ambient Intelligence.* London: Springer Science & Business Media.

5 The emotionally intelligent building

In the previous chapter we explored the concept of the intelligent building. This even extended to the idea of measuring a building's IQ or intelligence quotient. Is it possible or even desirable to assess a building's emotional intelligence? In order to do this, we need consensus about what is meant by an 'emotionally intelligent building'. Attempts to define intelligence itself have been fraught with conflict and disagreements amongst academics. Our preoccupation with intelligence has often been at the expense of emotional intelligence. As we shall see in this chapter, there are also disagreements about what the expression 'emotional intelligence' means. In this chapter we examine the popular and evidence-based ideas underlying emotional intelligence. We develop a basic framework based on the workspace as an artefact: an artefact with three key dimensions. Each of these dimensions are capable of harbouring emotional qualities.

Figure 5.1 Space efficiency versus delight.

> **Shedding light on things**
>
> It was a chance to catch up with the Chief Finance Officer, Martin, and find out his thoughts about the new building. We were attending the Town Hall event to celebrate the building's topping out. Both of us propped our elbows on the railings of the mezzanine floor, looking down at the assembly of the great and the good awaiting the address by the CEO. As facilities manager I knew this space was going to be something really special: not just to 'satisfy' people but to create a sense of 'delight'. Looking across the atrium, the opalescent afternoon sunshine infused the space. It created a sense of theatre for an office building that would otherwise be an efficient yet unremarkable space.
>
> "What's your verdict then, Martin? What do you think of it?" I asked.
>
> "It's pretty, but it comes at a cost! We were looking at a Utilisation Factor of 0.8 until the designer came up with this 'bright' idea of a spiritual hub. I've done the numbers on it. The atrium is costing us a further 2 per cent on construction and operating costs. The Utilisation Factor drops to 0.75. We won't be doing this again on our next building."
>
> I went quiet. I wasn't about to make him a convert. I just wished that we had some way of making all this show up on his spreadsheet.
>
> I couldn't help saying, "Yes, you won't be able to fit more desks in, but isn't it going to change the way we see things? Surely it will allow each of us to make sense of our part in the theatre of life?"
>
> Martin responded, "If you can't measure it, you can't manage it ... and you certainly can't cost it."
>
> I gave what I thought was a clever retort – a quote from Einstein: "Not everything that counts can be counted, and not everything that can be counted counts." But secretly I knew he was right. We have to find a way of measuring and managing this concept of 'spirituality' if we are to get Martin on board.

It's sometimes hard to put a name to it. But we know when it's not there. When we refer to people we might describe it as empathy or emotional intelligence. When we think of the workplace, we might describe it as heart, spirit or soul. When it's not there, we are all-too-familiar with the consequences. Let's take you back to 2013 when the CEO of Yahoo, Marissa Mayer, issued an edict banning employees from working 'remotely' (Arthur 2013). The human resources manager echoed this instruction: "... to become the absolute best place to work, communication and collaboration will be important, so we need to be working side-by-side. That is why it is critical that we are all present in our offices." But why do people choose to work remotely? One simple response might be 'because we can' – in other words, the office has nothing to offer that we cannot already get from the convenience of home working. Indeed, for the intelligent building the success of remote working might be seen as a measure of success. For an emotionally intelligent building, the opposite applies. Remote working misses out on something that the emotionally intelligent building has to offer.

Mayer explained just how important the office is as a communication tool: "Some of the best decisions and insights come from hallway and cafeteria discussions, meeting new people, and impromptu team meetings ... Speed and quality are often sacrificed when we work from home" (Arthur 2013). Do organisations have to rely on edicts to convince employees to show their face at work? How can an organisation provide a 'home from home' environment that isn't your home?

Meyer's strategy places the office building centre stage. But perhaps the reason for it is misplaced. Rather than 'speed and quality', shouldn't we be looking for that magnet that

44 *The emotionally intelligent building*

draws employees to work. Organisations grapple to understand why people are voting with their feet; why working from home appears so much more attractive. Yes, tasks are completed. But organisations are haemorrhaging social capital as employees continuously move in and out through the turnstile. In the following sections we explore what that elusive 'something' is that provides the glue for organisations: social intelligence and its modern conceptualisation, emotional intelligence.

Defining emotional intelligence

There's a medical condition that was identified in the middle of the last century. The condition is known as alexithymia. Whilst it may be little-known, it is suffered by 10 per cent of the population (Ruesch 1948, Maclean 1949). People suffering with alexithymia have difficulties identifying and describing their own emotions. Specifically, they have difficulties in regard to emotional awareness, social attachment and interpersonal relating. This difficulty in distinguishing and appreciating the emotions of others often leads to unempathetic and ineffective emotional responses in social settings. Perhaps you might recognise some of these characteristics in colleagues. We might say "He has sharp elbows" or "He leaves dead bodies in his path." However, we rarely see emotional shortcomings in ourselves. Emotional intelligence takes a different approach. Rather than 'fixing' a known mental condition, it considers the competence of the general population. It embraces the concept of positive psychology and the possibilities of enhancing our own emotional awareness.

Figure 5.2 The emotionally intelligent building (EQ) versus the intelligent building (IQ).

It has been claimed that people with an average IQ outperform those with the highest IQ 70 per cent of the time (Bradberry and Greaves 2009). And yet we place so much value on intelligence. Perhaps we should be looking elsewhere to understand what that special 'something' is that leads to personal success. Extending the idea to buildings, perhaps we have placed too much store on building intelligence. Perhaps like human intelligence, building intelligence is overrated and we should be looking for that critical factor that enables a building to fulfil its full potential.

The phrase 'emotional intelligence' (EQ as opposed to IQ) has been used to describe that special 'something' in each of us. It describes a largely intangible ability that we all have to manage behaviour, navigate social complexities and make social decisions. Time and again, emotional intelligence has proven to be a better predictor of performance. This explains the unprecedented interest shown by consultants, educationalists, researchers and the media. Some might say that it has become the victim of its own success, being defined in varied and often unscientific ways. Prior to any conception of emotional intelligence, two core intelligences were recognised. The first of these (verbal/propositional intelligence) deals with words and logic. The second of these (perceptual/organisational intelligence) deals with the arrangement and manipulation of objects in space. But as early as 1920, Thorndike (1920, p. 228) expressed ideas of a third missing intelligence which he described as 'social intelligence' that involved "the ability to understand and manage men and women, boys and girls – to act wisely in human relations". It was not until the latter part of the twentieth century that emotional intelligence (EQ) was coined as a phrase to describe this elusive third intelligence. It brought back to life some of the original ideas of Charles Darwin that some types of emotional information (for example, human facial expressions) are universal.

Can a workspace have emotional intelligence?

Is it possible for a workplace to have an emotional intelligence? Unless you're a devotee of panpsychism (the doctrine or belief that even inanimate objects have a level of consciousness), you might struggle with the idea of a workplace that actually 'feels'. However, we feel comfortable about the idea of an intelligent workspace. Some commentators have even gone so far as to attempt to measure workplace IQ as discussed in the previous chapter (Duffy 1985). In the following sections we explore the idea of emotional intelligence as a measurable phenomenon. Psychologists talk about tools and instruments as a way of probing and revealing this emerging concept. The competing definitions and tools have been widely criticised in terms of robustness and over-generalisation. Against this backdrop of controversy, what still stands up to scrutiny? Just as importantly, can we identify a measurement tool that captures the seemingly intangible characteristics of emotional intelligence in the physical environment? Is there an aspect of our workplace environment that brings out the best in us? Instead of looking at the emotional disabilities of one another, perhaps we should begin by acknowledging the emotionally disabling environments in which we work. In the following sections, two dominant models are described for evaluating emotional intelligence: (a) the ability model; and (b) mixed models (trait and abilities). We reflect on whether either model can be used to express the 'emotionally intelligent workspace'.

Ability model

The ability model (Mayer et al. 2008) thinks of emotional intelligence as a specific form of intelligence. We use a set of interrelated abilities to exercise our emotional intelligence. The proponents of the model suggest that "some individuals possess the ability to reason about and use emotions more effectively than others" (Mayer et al. 2008, p. 3). The model assumes that emotional intelligence overlaps with cognitive ability and reflects an additional facet of intelligence. They define emotional intelligence as "the ability to monitor one's own and others' feelings and emotions, to discriminate among them and to use this information to guide one's thinking and actions" (Salovey and Mayer 1990, p. 189). They use a Four-branch Model to represent these distinct abilities:

1 Managing emotions so as to attain specific goals.
2 Understanding emotions, emotional language and the signals conveyed by emotions.
3 Using emotions to facilitate thinking.
4 Perceiving emotions accurately in oneself and others.

When we think of a workplace environment, we can easily envisage how it can enhance or hinder each of these four abilities. Often the physical environment is intimately linked to job design, and both job design and the physical environment seem to get in the way of our emotional competence. The visibility of employees across an office space or the insecurity of employees arising from a hot-desk policy are two examples where emotions are managed or mismanaged in the pursuit of company goals. When it comes to understanding or conveying emotions, we are all-too-familiar with the misunderstandings that can occur at the reception desk. Only small adjustments to the receptionist's desk can make all the difference to the interaction (both cognitive and affective). We also know that a bright stimulating environment can dramatically affect our mood and motivation – thus facilitating our thinking process. When it comes to perceiving emotions, environmental factors such as lighting, noise and occupant density can significantly impact on our ability to detect emotional signals.

Measuring emotional intelligence using the ability model involves the use of ability-based scales involving eight tasks (known as the Mayer–Salovey–Caruso Emotional Intelligence Test or MSCEIT). One example of an Understanding Emotions task would be the following:

> What feeling, when intensified and coupled with a sense of injustice, is most likely to lead a person to experience anger? (a) frustration (b) guilt (c) fatigue.
> (Mayer et al. 2008, p. 507)

The degree of correctness would then be determined by the answers provided by a group of emotion experts (i.e. emotion researchers) or a sample from the population at large. The best answer to the question above would be (a) frustration, since frustration when intensified leads to anger. This type of scoring method is similar to that used to establish intelligence in classic intelligence tests.

Mixed models

An alternative model used to define emotional intelligence is the 'mixed model' – so-called because it includes both traits and abilities. Using this approach, emotional intelligence is "not classified as an intelligence but rather as a combination of intellect and various measures of personality and affect" (Joseph and Newman 2010, p. 55). One of the key proponents of the mixed model, Bar-On (2004), defines emotional intelligence as "an array of non-cognitive capabilities, competencies, and skills that influence one's ability to succeed in coping with environmental demands and pressures" (2004, p. 16). The mixed model has been the subject of criticism for two reasons: (1) emotional intelligence is based upon identifying all of those abilities that are not captured by cognitive ability; and (2) they are too redundant, reproducing many of the characteristics that represent personality traits. Critics of the method include Mayer *et al.* (2008, p. 505) who question the entire approach:

> A large number of personality traits are amassed, mixed in with a few socio-emotional abilities, and the model is called one of emotional intelligence or 'trait EI'. Generally speaking, these models include little or no justification for why certain traits are included and others are not, or why, for that matter, certain emotional abilities are included and others are not, except for an occasional mention that the attributes have been chosen because they are most likely to predict success.

Despite these criticisms, the mixed approach continues to receive attention because of its predictive power. The model used by Bar-On (2004) falls into this category of being a mixed approach. Key components of EQ include: (a) the ability to recognise, understand and express emotions and feelings; (b) the ability to understand how others feel and relate with them; (c) the ability to manage and control emotions; (d) the ability to manage change, adapt and solve problems of a personal and interpersonal nature; and (e) the ability to generate positive affect and be self-motivated.

To summarise, there are two distinct interpretations of emotional intelligence: (1) a theoretically narrow set of constructs concerned with the recognition and control of personal emotion (referred to as 'ability based' EQ); and (2) an umbrella term embracing a wide-range of constructs that have one thing in common: they are nothing to do with cognitive intelligence (this method is referred to as 'mixed based' EQ). In the remaining sections we examine whether a robust method might be used to evaluate the emotional intelligence of a work environment. We take a radical position, asserting that emotional dysfunction in the workplace is not the result of emotional inadequacies of employees. Instead, we suggest that workplaces themselves create emotional dysfunction. By understanding how work environments influence emotional well-being, we hope to define a preliminary framework to allow the evaluation of EQ for the emotional quotient of work environments.

Defining the emotionally intelligent workplace

Is the concept of an emotionally intelligent building a new idea? Evidence through history suggests that it is far from being a new idea. Indeed, it is entrenched in many cultural traditions. The term 'spirituality' or 'soul' are terms that have been used to characterise such spaces. As pointed out by Stokols (1990), the creation of spiritual spaces has a global heritage including the ancient tradition of Shintoism (Cali and Dougill 2012), which considers the arrangement of space in order to evoke sacred spirits; the long-standing Chinese practice of Feng Shui, which prescribes the selection and configuration of sites that are conducive to well-being (Bonaiuto et al. 2010); and the tradition of temple design in Hindu, Islamic and Christian traditions.

Has modern office design marginalised this idea of emotional empathy? In the pivotal paper by Stokols entitled *Instrumental and spiritual views of people–environment relations* (1990), the author challenged our blinkered behavioural approaches to building design. He questioned our preoccupation with the 'instrumental' approach to environmental design whereby the office environment is seen as simply a tool – a physical means for achieving key behavioural and economic goals. He suggested that the instrumental possibilities of space and physical settings neglected another less tangible but no less important characteristic of space – that of spirituality. In modern-day parlance, we might equally use the words empathy, mindfulness or emotional intelligence. From this perspective, the physical setting is seen not simply as a tool but as an end in itself – as a "context in which important human values can be cultivated" (Stokols 1990, p. 641). Table 5.1 shows the two contrasting approaches between the intelligent workspace (IQ) and the emotionally intelligent workspace (EQ) illustrating the tension between 'instrumentality' and 'spirituality'.

Stokols went on to argue that the instrumental perspective fell short when it came to understanding multifunctional settings:

> the potential for counter-productive programming may increase due to the diverse and sometimes competing interests among activities and occupants. In such situations, sophisticated programming and assessment techniques are required to identify the unique preferences of different user groups and the multiple symbolic meanings conveyed by the physical environment.
>
> (Stokols 1990, p. 644)

Table 5.1 Contrasting assumptions between the intelligent workspace (IQ) and the emotionally intelligent workspace (EQ)

Intelligent workspace (IQ)	Emotionally intelligent workspace (EQ)
Instrumental view whereby settings are simply tools for promoting behavioural and economic goals	Emotionally intelligent (spiritual) view whereby settings enable human values to be cultivated
Quality is defined in terms of behavioural, comfort and health criteria	Quality is defined in terms of the richness of the psychological and sociocultural context
Emphasis on the material and technological features of the workspace	Emphasis on the symbolic and affective features of the workspace
Technological connectedness	Human connectedness
Doing	Being
Environmental design standards using prototypes for universal standards	Creation of customised designs to meet the unique requirements of specific individuals, groups and cultures
Emphasis on the creation of distinct separation between public- and private-life domains	Recognition of blurred boundaries between public life and private life; employee and contractor (coworking)
Standardisation and modularisation are the dominant design approaches	Uniqueness and customisation arise from contextualised and empathetic design
Tangible (measurable)	Intangible (difficult to measure)

In contemporary multiple-activity work settings, the wide range of individual and group activities invariably require a correspondingly diverse set of instrumental and symbolic interpretations.

On the same theme, Franck (1987, p. 65) highlighted the re-emergence of the symbolic and spiritual aspects of environmental design when she contested that designers "are becoming increasingly interested in history, culture, myth and meaning". Other contemporaries called for "a new sensibility in design in which human activity, human feeling, colour and light together create an ordinary human sweetness, something almost entirely missing from the works of this century" (Alexander et al. 1987, p. 129).

Making sense of it all

> Harry muscled into the busy elevator. It was six months since they had moved into the new corporate office building. The relocation was supposed to herald a new era. But the off-hand comments in the elevator suggested a new name for the building had gained currency. People called it the 'Teflon' building because it was impossible to leave a personal mark on the place. Worse still, it was difficult to see beyond the building's public image: to actually make sense of this organisation based on what could be seen around you. For Harry, who was a newcomer, the building really didn't speak to him.
>
> Bunched up in the corner of the elevator, he was forced to look at himself in the mirror. He saw the toothpaste that had stubbornly remained on his chin all morning, unbeknown to him. At least this building provided some sort of feedback, if only about his appearance. He had gone all morning without anyone telling him! If only this office environment was able to reveal how this organisation worked. For now, it remained an impenetrable mystery.

The type of trauma that Harry describes as a recent employee is not uncommon for many people. Alvin Toffler referred to it as 'future shock', which he described as "a state of distress or disorientation due to rapid social or technological change" (1970, p. 2). How often are we as designers of work environments responsible for subjecting people to this trauma? The absence of cues around us can make it difficult to 'read' an organisation.

An emotionally intelligent building should provide 'intelligence' in the form of feedback or cues. This process can be described as 'sensemaking', a term coined by Karl E. Weick in the 1970s that referred to "the ongoing retrospective development of plausible images that rationalise what people are doing" (Weick et al. 2005, p. 409).

Some of the properties of a sensemaking environment include:

- the ability to establish who an individual thinks they are in their context (Pratt 2000);
- the ability for an environment to grab attention and interrupt – creating an opportunity for reflection (Dunford and Jones 2000);
- the ability for workplace environments to convey narratives and dialogues (i.e. tell a story) such that people can think, organise and control their experiences (Currie and Brown 2003);
- the ability of an environment to provide ongoing feedback that will enable individuals to project themselves onto this environment and observe the consequences they learn about the identities and accuracy of their accounts of the world (Thurlow et al. 2010).

50 *The emotionally intelligent building*

Much of the research on sensemaking makes use of written and documented narratives. It forms part of a tradition of qualitative organisational research. There is much less written about the potential of artefacts and environments to tell a story. And yet, the potential of physical space to enable sensemaking is far greater – first, it is omnipresent, and second, it infuses our consciousness on a daily basis without us even knowing it.

Sensemaking in the organisational literature also pays little attention to emotions. It has been argued that "despite the widespread evidence that emotion is an integral part of sensemaking […] relatively little theory has been developed that explicitly identifies the roles that emotion plays in sensemaking and its impact on sensemaking processes" (Maitlis et al. 2013, p. 223).

The three dimensions of workspaces

Conflicting perspectives

It had become a major cause of disagreement and threatened to undermine the whole office initiative. Chiara was the facilities manager and had become enraged that the interior designer, Petra, had decided to dispense with the skirting boards in the plush new C-suite on the seventh floor.

"How can we possibly do without the skirting boards? Doesn't Petra understand why they are there?" remarked Chiara.

"They're an eyesore! They are an anachronism left over from the days when gentlemen wore spurs and the wall and wallcoverings were at risk from a stray boot! I want to go for the minimalist approach", Petra responded.

"You're wrong. Even back home in Italy where your minimalist approach is so popular, we use skirting boards. There we call them *battiscopa* which simply means 'where the broom hits'. Can't you see the problems involved with cleaning this space?" Chiara asked, feeling overlooked.

"I guess it gives the CEO something to kick when he sees the drop in share price", remarked the CFO, rather churlishly.

"I'm not interested in your cleaning problems", Petra remarked, ignoring the throwaway remark of the CFO. "I want a space where we can showcase our company achievements. How to manage without skirting boards is your problem!"

At this point, the head of corporate branding, Colin, interjected.

"Ahhh … You mean the base board. That's what we call it back in the US. Couldn't we use the colour puce to paint the base boards? It's our corporate colour! Everyone would connect this with our bestselling product."

Chiara rolled her eyes. Before, she only envisaged snagging and maintenance difficulties. Colin's colour scheme would surely contravene disability legislation.

This story illustrates how different professionals value or neglect different aspects of the same 'artefact' (we can think of the workspace and the objects within it as artefacts since they are man-made). Although this scenario is fictitious, it portrays the kind of dialogue that happens every day as designers grapple with solutions that meet the needs of different stakeholders. Clearly, we can see in the heated discussion that the 'instrumental' views of the facilities manager are at first sight the most compelling. We can see that through this lens we have an objective view based on measurable performance criteria and existing design standards. But Petra's concerns as the designer focus on the aesthetic impact of the skirting boards and her emotional response. Whilst her objections might seem reasonable, she stands on stony ground as there is a notable absence of objective measures to defend her position. Colin sees the branding opportunities and the messages conveyed by the colour of the contentious skirting boards. This points to a third dimension of artefacts: their symbology. Taken together we can see the three dimensions – instrumentality, aesthetics and symbology – present in almost every artefact. The following sections explore the possibility of using these three dimensions to characterise emotional intelligence.

The green bus

This story begins with the seemingly innocuous decision to paint a fleet of buses green. In October 1999, the public transportation company of Israel decided to paint all of their 3,800 buses green. This decision didn't involve any pretesting. The intention was to promote the organisation as 'natural' and 'environmental'. But the decision backfired. It resulted in unexpected and highly charged reactions that were entirely unforeseen.

This initiative prompted a study by Rafaeli and Vilnai-Yavetz (2004a) from the Technion University in Israel who remarked on the serendipity of the situation:

> That the colour was introduced without pretesting provided an unusual opportunity to examine sensemaking of an artefact that was introduced in a real-life setting but was not – as often happens – shaped to meet expectation of various constituents.
> (Rafaeli and Vilnai-Yavetz 2004a, p. 674)

The study itself sought to achieve two things:

1 to validate the idea that three dimensions (instrumental, aesthetic and symbolic) underlie sensemaking of organisational artefacts; and
2 to examine the emotion that enters this sense making process.

The study involved employees of the company ($n = 32$), marketing professionals ($n = 15$), design professionals ($n = 12$), technical professionals ($n = 45$) and passengers or potential passengers ($n = 70$). The artefact of interest was the bus itself. To all of the stakeholders, the bus fleet represented the organisation in some way. Questions to elicit emotional responses were employed, such as:

> "What function do you see the green colour serves? Why? If you were to redesign the buses, what would you suggest? Why?"
> (Rafaeli and Vilnai-Yavetz 2004a, p. 675)

A small amount of variation with the questionnaire protocol was used in order to accommodate the expertise of particular respondents. Engineers were asked more in-depth questions about usability; designers about style; and marketing professionals about symbolism.

Some of the positive emotional terms that emerged included joyful, calm, restful, good, happy, arousing and pleasant. Negative emotional responses included revulsion, disgust, fear, melancholy, stress, worry and irritation. What became evident in the study was just how varied people's emotional responses were to a green bus. The same artefact could produce very different and often extreme negative emotions. Environmentalism is only one of the associations that a green bus evoked. At the negative end of the spectrum some respondents associated the green bus with a terrorist group that favoured the colour green. One of the drivers responded:

> "It is very negative. The terrorists may find the colour nice. Ask them, and they will like it. But for everyone here it is awful!"
>
> (Rafaeli and Vilnai-Yavetz 2004a, p. 681)

Clearly when it comes to symbolism, a seemingly innocuous and simple design decision such as colour can evoke many different feelings. As such, the study suggests that designers should think beyond simply the 'intended qualities' and consider other unforeseen emotional responses.

The findings of the research identified three key points:

1 sensemaking of artefact (objects) appeared to emerge along three dimensions (instrumental, aesthetic and symbolic);
2 this sensemaking process produced an emotional response to the artefact;
3 this sensemaking process also evoked emotion towards the organisation.

The study by Rafaeli and Vilnai-Yavetz (2004a) has prompted other studies that have used the same three dimensions to capture emotional response. Most notable is the follow-up study on office design by the same researchers (Vilnai-Yavetz et al. 2005) that demonstrated how the three separate dimensions of instrumentality, aesthetics and symbolism relate to the spatial artefact we know as 'the office'. Instrumentality provided an indicator of employee satisfaction and effectiveness, whereas aesthetics was found to be only related to satisfaction, whilst symbolism was not related to satisfaction or effectiveness. The framework was also endorsed by Elsbach and Bechky who illustrated in *California Management Review* (2007) how decision-making in office design could be improved using this approach.

Putting the pieces together

We now go to the heart of this book's thesis: that the office and the spaces and elements that make it up can be described as artefacts, each artefact having three dimensions – instrumental, aesthetic and symbolic. Emotion plays a part in each of these dimensions but the mechanisms that evoke such emotions are dissimilar for each. This is where things get interesting. Applying the wrong appraisal and feedback mechanisms (e.g. performance standards or satisfaction surveys) creates an inevitable bias so that the more easily measured instrumental dimension is overvalued at the expense of aesthetic and symbolic characteristics. In the following three chapters we explore three very different mechanisms by which emotions are evoked, each tied to one of the three dimensions illustrated in Figure 5.3.

Emotional mechanisms

Research by (Rafaeli and Vilnai-Yavetz 2004b) suggests that the emotional triggers that arise from the instrumental, aesthetic and symbolic assets are fundamentally different. Figure 5.4 illustrates the differing mechanisms associated with the three dimensions.

The emotionally intelligent building 53

Figure 5.3 The dimensions of an artefact (instrumental, aesthetic and symbolic).

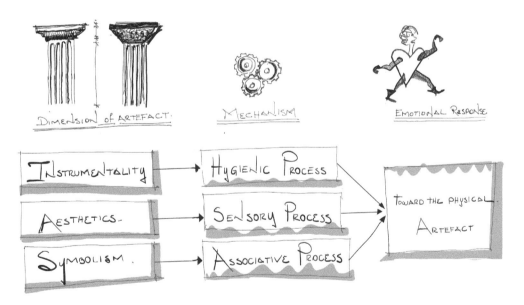

Figure 5.4 The different mechanisms influencing instrumentality, aesthetics and symbolism.
Source: Rafaeli and Vilnai-Yavetz 2004b.

- *Instrumental:* Emotions arising from the functional or instrumental dimension are tied to levels of dissatisfaction. In other words, emotions are linked to hygiene factors. Eliminating dissatisfaction is always the goal, but going beyond this does not necessarily lead to satisfaction. It leads to a design philosophy of getting it 'right' and 'fixing it': eliminating negative emotions.

- *Aesthetic:* Emotions associated with the aesthetic dimension arise from the senses – we feel emotions as a result of direct interaction with minimal cognitive engagement. It is an immediate and spontaneous response. It leads to a design philosophy of 'creativity, play and interaction': stimulating positive emotions.
- *Symbolic:* Emotions arising from the symbolic dimension involve an interaction between the senses, cognition and memory. Seeing a place or an object triggers a memory or an association. We often use artefacts to trigger associations we want others to have about us. At a corporate level this might be described as branding. At an individual level, it might be described as personalisation. The resulting emotions typically reflect variations in culture and past experiences. It leads to a design philosophy of 'creating and communicating the right message'.

We might think of items in the office landscape as serving either as a tool, as something that simply adds to its appeal, or as a sensemaking device that enables us to interpret our surroundings. In practice, these three dimensions are not mutually exclusive. As designers and managers, we continually exploit these different aspects.

Creating the emotional palette

Extending the ideas of (Rafaeli and Vilnai-Yavetz 2004b), we could devise an emotional design palette that could be used to direct a designer's thinking. This is illustrated in Table 5.2.

We can think of the painter whose palette depends upon three primary colours: red, yellow and blue. From these colours the painter is able to mix a range of secondary colours that may give rise to a feeling of warmth or coldness, excitement or anxiety. In just the same way, the three dimensions of any artefact are interrelated and can lead to a range of emotional responses (see Figure 5.5).

Most organisations focus on change and output. They are future oriented. The instrumentality of the workspace enables employees to 'get the job done'. This focus on 'doing' reflects a utilitarian view of the office. However, for a growing number of creative organisations, the workplace 'experience' needs to enliven their senses. This shifts the timeline towards the present – towards the aesthetic experience. The focus is on 'being'. Moving

Table 5.2 Workspace characteristics expressed as instrumental, aesthetic and instrumental dimensions

Dimension	Symbolic	Aesthetic	Instrumental
Drivers	Past driven	Present driven	Future driven
Activity	Belong	Be	Do
Mechanism	Associative	Experiential	Hygiene
Relationship with the workspace	...Attach to others	...Interact with	...Use
Activity setting	Team working	Deep working	Private working
Connectivity	People connected	Disconnected	IT connected
Preference	Culture	Individuality	Expertise
+ve emotions	Belonging	Pleasing, arousing	Neutral, indifferent
-ve emotions	Alienation, anxiety, disgust	Boredom, annoyance, disgust	Stress, anger

Source: Partly derived from work by Rafaeli and Vilnai-Yavetz 2004b.

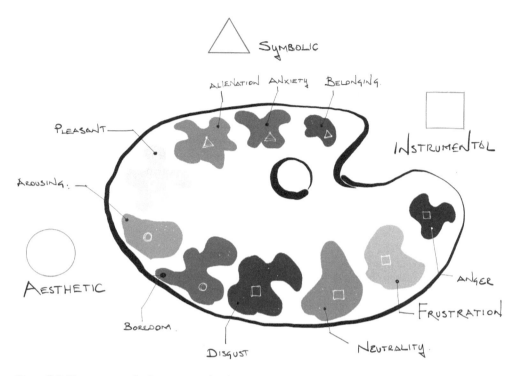

Figure 5.5 Designing with the emotional palette.

further along the timeline, we address the past. Where do we come from? What do we represent? Who do we belong to? The focus becomes 'belonging' as part of a social sphere. We look to the workspace for cues to enable us to feel grounded.

A cynic might argue that employees only get paid for output: that the workspace is simply a set of tools – "we should only concern ourselves with the future". But in order to be productive, having a sense of the present and the past is paramount. As the microbiologist and novelist Renee Dubos argued, we should engage with the world having a "… reverence for the past, love for the present, and hope for the future" (Dubos, 1965, p. 279). The workspace has the potential to provide this motive or driving force.

An emotionally intelligent work setting

Our intention in the following chapters is to present a framework for achieving the right 'office chemistry'. Central to the idea of an emotionally intelligent work setting is its fit with an individual and the organisation. The starting definition is: the sum of emotional feedback that can be perceived, discovered, or learned as a result of being in and engaging with a work setting – both individually and as groups. This intelligence enables sensemaking as a result of interaction with space and artefacts within the space, using one of three dimensions: instrumentality, symbolism or aesthetics.

Summary

The following three chapters explore the instrumental, symbolic and aesthetic dimensions. In all three chapters we examine the science and evidence driving workplace design. Rather than seeing the instrumental dimension as being a purely cognitive feature of the office, the next chapter illustrates how the emotional facet of negative emotions can intervene. It illustrates how recent innovations such as 'nudge' are stimulating individual and organisational performance without recourse to technology. In the subsequent chapter, we examine how the second dimension, that of aesthetics, can support the individual and the organisation. In the chapter on symbolism, we uncover how it provides a means of self and organisational expression. Each of the chapters contains one or more case studies to illustrate how emotional intelligence can be harnessed – supporting the doing, being and belonging aspects of a workplace environment.

References

Alexander, C., Anninou, A., Black, G. and Rheinfrank, J., 1987. Towards a personal workplace. *Architectural Record Interiors*, February, 129–141.

Arthur, C., 2013. Yahoo chief bans working from home. *Guardian*, 25 February.

Bar-On, R., 2004. The Bar-On Emotional Quotient Inventory (EQ-i): Rationale, description and summary of psychometric properties. In: *Measuring Emotional Intelligence: Common Ground and Controversy*. Hauppauge, NY: Nova Science Publishers, 115–145.

Bonaiuto, M., Bilotta, E. and Stolfa, A., 2010. 'Feng Shui' and environmental psychology: A critical comparison. *Journal of Architectural and Planning Research*, 27 (1), 23–34.

Bradberry, T. and Greaves, J., 2009. *Emotional Intelligence 2.0*. San Diego, CA: TalentSmart.

Cali, J. and Dougill, J., 2012. *Shinto Shrines: A Guide to the Sacred Sites of Japan's Ancient Religion*. Hawai'i: University of Hawai'i Press.

Currie, G. and Brown, A.D., 2003. A narratological approach to understanding processes of organizing in a UK hospital. *Human Relations*, 56 (5), 563–586.

Dubos, R., 1965. *Man Adapting*. New Haven, CT: Yale University Press.

Duffy, F., 1985. ORBIT-2: Know your building's IQ. *Facilities*, 3 (12), 12–15.

Dunford, R. and Jones, D., 2000. Narrative in strategic change. *Human Relations*, 53 (9), 1207–1226.

Elsbach, K.D. and Bechky, B.A., 2007. It's more than a desk: Working smarter through leveraged office design. *California Management Review*, 49 (2), 80–101.

Franck, K.A., 1987. Phenomenology, positivism and empiricism as research strategies in environment-behavior research and in design. In: *Advances in Environment, Behavior and Design, Vol. 1*. New York: Plenum Press, 59–67.

Joseph, D.L. and Newman, D.A., 2010. Emotional intelligence: An integrative meta-analysis and cascading model. *Journal of Applied Psychology*, 95 (1), 54–78.

Maclean, P.D., 1949. Psychosomatic disease and the 'visceral brain': Recent developments bearing on the Papez theory of emotion. *Psychosomatic Medicine*, 11, 338–353.

Maitlis, S., Vogus, T.J. and Lawrence, T.B., 2013. Sensemaking and emotion in organizations. *Organizational Psychology Review*, 3 (3), 222–247.

Mayer, J.D., Salovey, P. and Caruso, D.R., 2008. Emotional intelligence: New ability or eclectic traits? *American Psychologist*, 63 (6), 503–517.

Pratt, M.G., 2000. The good, the bad and the ambivalent: Managing identification among Amway distributors. *Administrative Science Quarterly*, 45 (3), 456–493.

Rafaeli, A. and Vilnai-Yavetz, I., 2004a. Emotion as a connection of physical artifacts and organizations. *Organization Science*, 15 (6), 671–686.

Rafaeli, A. and Vilnai-Yavetz, I., 2004b. Instrumentality, aesthetics and symbolism of physical artifacts as triggers of emotion. *Theoretical Issues in Ergonomics Science*, 5 (1), 91–112.

Ruesch, J., 1948. The infantile personality: The core problem of psychosomatic medicine. *Psychosomatic Medicine*, 10, 134–144.
Salovey, P., and Mayer, J. D. (1990). Emotional intelligence. *Imagination, Cognition, and Personality*, 9, 185–211.
Stokols, D., 1990. Instrumental and spiritual views of people–environment relations. *American Psychologist*, 45 (5), 641–646.
Thorndike, E.L., 1920. Intelligence and its uses. *Harper's Magazine*, 140, 227–235.
Thurlow, A., Mills, A.J. and Helms Mills, J., 2010. Making sense of sensemaking: The critical sensemaking approach. *Qualitative Research in Organizations and Management: An International Journal*, 5 (2), 182–195.
Toffler, A., 1970. *Future Shock*. London: Bantam.
Vilnai-Yavetz, I., Rafaeli, A. and Yaacov, C.S., 2005. Instrumentality, aesthetics and symbolism of office design. *Environment and Behavior*, 37 (4), 533–551.
Weick, K.E., Sutcliffe, K.M. and Obstfeld, D., 2005. Organizing and the process of sensemaking. *Organization Science*, 16 (4), 409–421.

6 Emotion and the instrumental workspace

What place does emotion have in a world preoccupied with utility and new technology? The instrumental workplace provides a tool attuned to organisational and personal efficiency. It represents a focus on ends rather than means. It's all about doing, changing and producing – and is future oriented. Nevertheless, even in a world of technology, emotion plays its part. Often our efforts in the instrumental workplace revolve around firefighting and quashing negative emotions – fixing technology that doesn't work or people that encroach on our space.

But isn't the workplace more than simply what we get out of it? Just like the consumer industry, workplace designers now recognise how interaction can actually be pleasurable. We are also finding ways to influence people's workplace choices so that 'hardwired' routine practices can be replaced by better ways of doing things. In this chapter we will see innovative ways in which the affective aspects of workplace are bringing about positive organisational outcomes.

Understanding the instrumental perspective

"I don't care about the 'fluffy' stuff – I just want a workspace that allows me to get on with the job." How often do we hear these sentiments? They express our common frustration with technology, access to resources and the upheaval of sharing. Emotion is as much part of the instrumental worldview as the aesthetic or symbolic view. But our focus on workplaces as simply providing a means to an end brings about particular types of emotion. At one end of the spectrum, when things don't work, we experience frustration or even anger. At the other end, when things are resolved, frustrations disappear. If we view a workspace as 'working' or 'not working', we confine ourselves to what has previously been described as hygiene factors. As a facilities manager, we are in a state of continual alert, looking for signs of dissatisfaction. The best that we can hope for is to extinguish people's discontent. This is consistent with the Two-Factor Theory of Herzberg who argued that remedying the causes of dissatisfaction does not lead to satisfaction. Nor will adding satisfiers eliminate dissatisfaction. That is because the opposite of satisfaction is 'no satisfaction', and the opposite of dissatisfaction is 'no dissatisfaction'.

> So often it becomes the functionality ... as long as we can get the job done. And we don't worry about how much of themselves they might be bringing to the work ... or not bringing to the work.
>
> *Interview with consultant in community development*

The non-territorial office

'One size fits all' remains the mantra for many office solutions. The standardisation and miniaturisation of workstations has historically been driven by real estate concerns to reduce occupancy costs, reconfiguration costs and churn costs. For workers, it offers the prospect of a 'work anywhere' agenda. Interchangeability of workstations combined with mobility using computers, tablets and smart phones has untethered the office worker. Organisations have also recognised the organisational benefits of teleworking (working remotely) or using the physical office space more flexibly. The most commonplace form of flexible office involves the use of 'non-territorial' workplaces or what we might describe as hot-desking. In such a system, "staff have no fixed personal workspace and use any available desk as needed" (Elsbach 2003, p. 622). The term hot-desking originates from the expression 'hot bunking' or 'hot racking' – a practice widely used by submariners when "assigning more than one crew member to a bed or 'rack' to reduce berthing (sleeping) space" (Jackson 2010, p. 15).

Whilst organisations might eulogise about the benefits of hot-desking, office workers have given it a mixed reception. A study of 1,000 workers in New Zealand (Morrison and Macky 2017) sought to investigate the 'demands [–] and the resources [+]' arising from a hot-desking strategy using the job demands-resources (JD-R) model (Demerouti and Bakker 2007). Most of the benefits (resources) of hot-desking accrue to the organisation as a whole. These included projecting an image of being modern and forward thinking, improving flexibility in the use of the physical space, creating closer working relationships, high productivity, increased networking opportunities, and more easily exchanged knowledge and skills.

When we look at the dis-benefits (demands) of hot-desking, the prevalence of emotions in the workplace become evident. The study by Morrison and Macky (2017) identified issues related to distractions, uncooperative behaviours, distrust and negative relationships. People were found to routinely mark and protect their workspace because it was seen as an essential part of the work experience. They found that the hot-desking condition increased distrust and undermined interpersonal relationships between employees compared with having one's own office or sharing with one or two others. Unsurprisingly, the levels of distraction reported in a shared office environment was significantly worse than working at home or on the road. There was also evidence of uncooperative behaviour. Similar findings (Brown 2009) suggest that people are expending a great deal of energy in order to assert their ownership, including squatting in offices that were meant to be vacant, reconstructing a specific area or territory for the group, displaying identity markers, or engaging in antisocial behaviours. These included saving spaces for others, not letting people use particular workstations or fixing pictures onto their equipment.

> In my previous role I moved out of my office to sit with my team. But that space was very artificial. We were in an in internal pod space which I found really challenging. It was very much set up in rows rather than pod style. It wasn't conducive to really good sharing and interaction – it felt very call-centre like.
>
> *Interview with consultant in start-up incubator space*

How do we overcome the evident problems in hot-desking? After all, the economic argument for adopting flexible 'non-assigned' offices are compelling. It is increasingly difficult to justify dedicated workspaces when an office worker uses several work locations both within the office and externally. The low occupancy levels of dedicated spaces can also remove any vibrancy. To address this challenge, we need to understand the elusive nature of emotion. In terms of getting the job done (instrumentality), how do we create spaces in which sharing is seen as a positive asset.

Laying down the law

We can think of an office as a discrete artefact in just the same way as the green bus that was discussed in the previous chapter. But an office represents much more than a simple configuration of materials and surfaces. A much greater challenge for the designer is understanding the office as a social system. Much of the tension that exists in the modern office environment arises from worker cooperation – and perhaps the lack of it. How many workplaces display this sign in the kitchenette: "Your mother doesn't work here – clean up your own mess."

Many commentators believe that with the advent of open plan sharing environments, the adoption of formal 'office etiquette' is necessary. One area that has attracted some interest is 'lean thinking'. It seems that we expend much of our working time making unnecessary trips, looking for documents or other resources. If we were to embrace a lean thinking approach we would know where to find the stapler or get our hands on a crucial document. In a shared environment, the problem of finding things is magnified. 'Waste' is also evident in travel times between workplaces and accessing resources. The supposition in the lean office is that waste is a widespread problem that can be remedied by applying five key principles (5S) developed in Japan:

- sort (*seiri*);
- set in order (*seiton*);
- shine (*seiso*);
- standardise (*seiketsu*);
- sustain (*shitsuke*).

The 5S system was a key element of the just-in-time manufacturing system. Its purpose was to maintain a neat and clutter free manufacturing environment. In manufacturing environments, clutter and disorder can have major safety implications. Imagine a misplaced spanner in an aircraft fuselage! In the office environment, the main argument for introducing a 5S system is to reduce waste (and secondarily to reduce conflict in a shared environment). Let's take each of the 5S items in turn.

1. *Sort:* What do I need to do my job? By asking this question, we start to identify those items in our workspace that add value to our daily routines. If it does not add value, it is taking up valuable real estate on our desk. This first stage will prompt us to sort/purge documents around us, being mindful of a company's retention policies.
2. *Set in order (or straighten):* Where does this belong? This second stage involves finding a 'home' for documents and office equipment so that you can easily retrieve it at a later date. True enthusiasts of 5S use labels liberally and even use outlines of office devices such as phones and laptops using colour-coded electrical tape!
3. *Shine (or sweep):* Should I clean it? Maintaining the equipment (e.g. cleaning a keyboard or stocking printer paper) not only makes equipment more serviceable, it also addresses issues of hygiene, particularly in a shared environment.

4 *Standardise*: Is there an office procedure? The idea here is that there is an agreed way of doing things – a particular method for filing systems, an agreed method for colour coding equipment, or perhaps a shared understanding of a hot-desking policy.
5 *Sustain*: How do I keep this going? The sustain step involves incorporating each of the other steps into a daily routine. It is designed to prevent the 'put things away until tomorrow' mentality.

In a sense, the 5S methodology is an example of many similar 'office etiquettes' that are routinely introduced in modern office environments. The idea of introducing 'discipline' evokes extreme emotional responses by users in the workplace, as illustrated in the following posts from a web forum (Does 5S make sense in an office environment? 2012).

> This 5S system is a recipe for disaster if imposed on a software development team. Yes, it sounds truly awful! I want my workplace like I'm used to, not as some 'standard' says it should be.

> 5S sounds great for 'interchangeable' people (e.g. factory assembly-line), but what happens when one of your office workers needs a different chair for his bad back, another needs a different resolution on the monitor to see, and the third is left-handed and needs to invert everything?

> I disagree entirely with the idea. When I do creative work, that's inherently messy and a requirement to keep everything spit-polished is not only ridiculous, it is counterproductive. I file visually. I can always find something in the stack on my desk but if it goes into a file/folder it is lost forever. It is wasteful of time to be forever cleaning when I could be working. Not everyone works well in neat conditions. I find them distracting and uncomfortable and I can't find what I need.

As we move towards shared environments, the need for mutual cooperation becomes evident. But what role does the designer or facilities manager have in this world of conflict and ambiguous boundaries between employees? Can we rely on office etiquette, or do less intrusive options exist? Getting a good fit between employee, job requirements and resources is a first step. But is fitting the 'type of worker' to the workstyle sufficient? Should we also consider the individual and group emotions that ensue in the shared environment?

Fitting the person to the workplace

How do we go about fitting the person to the workplace? It seems that one size does not fit all and the advent of flexible working has ushered in multiple options. We can start by addressing three simple questions:

- What type of work does the office worker undertake? (instrumental)
- What group or organisational style do they want to fit into? (symbolic or social)
- What are their preferred ways of working? (aesthetic or individual)

The report *Working Without Walls* (Allen 2004) and its successor *Working Beyond Walls* (Hardy 2008) have been highly influential in transforming UK government buildings based on 'fit'. An instrumental approach to fitting a person to a workstyle is proposed in the reports. Through a knowledge of work category, this mapping process allows a particular workstyle characteristic to be identified. In other words, 'What type of worker am I?' and, more importantly, 'What type of resources will I need?'

- If I am a 'process worker' or I am a team anchor then I would be described as having a 'resident' workstyle. Having your own desk and access to physical assets and internal interaction are paramount.
- If I am an externally mobile employee, having access to my own desk is not necessary. Use of a shared desk and access to mobile ICT enables me to become a 'nomadic' worker.

This approach of fitting a person to a workstyle based on job function satisfies the resource requirements of individuals by prescribing tools that are suited to the job.

Internally mobile?

HUMAN RESOURCES: I've talked with your line manager and I understand that you have been designated as a nomad/traveller.
NEW OFFICE WORKER: Well, I'm not sure what that means. I do have a tendency to go where the urge takes me.
HUMAN RESOURCES: Your designation as a nomad dictates what office workstyle you will adopt and the resources you will use.
NEW OFFICE WORKER: Does it mean that I get a campervan?
HUMAN RESOURCES: You won't have your own office desk. You will be in a shared office and external physical interaction will be prioritised over internal physical interaction. You will have minimal dependency on office systems but we need to set you up with the latest mobile ICT.
NEW OFFICE WORKER: I'm not sure what you mean by prioritising my external physical interaction, but let's wait and see.
HUMAN RESOURCES: It says here that you're not internally mobile...
NEW OFFICE WORKER: Hmm ... that sounds serious?

This scenario above might seem flippant but it characterises many people's response to being pigeonholed. We prefer to make our own judgements about workstyle, even if it is not the most effective. What is important is that we have made our own choice. Set against this are the organisational concerns about security, productivity and waste. The tensions between a mechanistic (prescribed) and an individualistic (elective) approach are most apparent in shared working environments.

User-centred design

Sometimes it is about getting the workspace design right in the first place. User-centred design (Eason 1995) seeks to close the gap in understanding between users and designers. This is achieved using a variety of research and design techniques. These include both 'investigative' (surveys and interviews) as well as 'generative' (brainstorming) tools and techniques. This usually follows the sequence:

> Understand context of use ☞ Specify user requirements ☞ Design solutions ☞ Evaluate against requirements

This approach has proven invaluable in incorporating complex user requirements in modern workplace environments. It has given rise to systematised methods such as post-occupancy evaluation (POE) that have consistently demonstrated the importance of user feedback (Preiser et al. 2015, Hay et al. 2018).

Nomad or vagrant?

Talk about being a 'nomad' and we are invited into a world of romantic, empowered or even maverick working. In reality, employees have the opposite experience. Mobile workers often feel dispossessed, no longer owning their own workplace. Hirst (2011, p. 768) expresses her concerns over the use of the 'nomad' metaphor:

> The 'nomad' metaphor is highly pervasive and extends beyond the telework literature … We appear, therefore, to be undergoing a significant shift in our relationship with the cities and workplaces we use in everyday life, in which we are increasingly 'hurried on' in order to allow the next consumer or worker to use the space.

In her ethnographic study of hot-desking environments, Hirst (2011) concludes that a new social structure has emerged as a result of official requirements for mobility. In this structure, which involves a loss of everyday ownership of the workspace, tensions have arisen between 'settlers' and hot-desking nomads, with some employees becoming dispossessed. Rather than referring to 'residents' or 'nomads', her ethnographic study spoke of two unofficial worker types:

- 'settlers' – referring to employees that 'own' particular desks enabling them to initiate and maintain relationships with colleagues as well as express their own identities;
- 'vagrants' – referring to the nomadic worker, being obliged to unpack and repack work and reinvest in a constantly changing social environment.

This dual culture can be part of the space planning process or emerge as a result of territorial claims of ownership. However, this two-stream workspace does not always bring about advantages for settlers. In the following section we examine one major concern for settlers – or, indeed, any worker who habitually works in a stationary sitting position: the ill effects of being seated for too long. How do we deter people from adopting a static sedentary workstyle? In the subsequent discussion, we identify the significant role played by emotions in achieving instrumental goals in the workplace. More specifically, we look at 'nudge' as a new concept that relies on emotion and behaviour change.

Replacing physical walls with psychological walls

Organisations have been busy tearing down walls with the introduction of open plan offices to replace conventional office space. As noted in the report *Working Without Walls* (Allen 2004, p. 17), the refurbishment of 1 Horse Guards Road in 2002 for the UK Treasury meant:

> More than seven miles of internal walls were removed to allow all staff to be accommodated in either perimeter team spaces or larger open plan areas around the internal courtyard.

64 *Emotion and the instrumental workspace*

The same process is being followed globally in order to harness the benefits of shared working environments. But the story does not end there: physical walls are being replaced by psychological walls – social constructs are superseding physical constructs. Employees are having to become more 'informed clients' as they negotiate the increasingly 'nuanced' world around them. Designers seek to employ more subtle interventions

Figure 6.1 Open plan design office.

that no longer rely on physical boundaries. Sensemaking, prompts and nudges have replaced signposts and corridors. Users increasingly become co-designers (Sanders and Stappers 2008).

The emergence of Activity Based Working (ABW)

One major innovation in workplace design promises to overcome many of the limitations of sedentary behaviour in the office. The concept of Activity Based Working (ABW) has its origins in a seminal paper written by Stone and Luchetti (1985). In their paper entitled *Your office is where you are*, they envisaged a workplace radically different from the office environment that prevailed at the time. They raised the question:

> How can managers enjoy the communication and participation advantages of the open office as well as the quiet and privacy of the closed office? How can they wander about the building without losing the convenience of being at their own desk?
> (Stone and Luchetti 1985, p. 102)

In the mid-1980s, the authors observed the clatter and disruption in the technology-laden open plan environment – which undermined work that increasingly required intense concentration. At one corporate, the authors noted that the "corporate nurse stocks earplugs". Today, we have replaced the earplugs with headphones. But the solution proposed by Stone and Luchetti (1985) was ahead of its time, predating (Wi-Fi) technology by some 20 years. Their three basic assumptions were:

- We use the ad hoc work team as well as the individual employees as our reference point.
- We assume that introducing office automation requires more than providing hidden tracks for cables. It requires managers to rethink how information and people flow in an office and adapt to what we call 'activity settings'.
- For many modern offices 'position' no longer means 'place'.

(Stone and Luchetti 1985, p. 103)

The premise of the 'activity settings' approach was that an all-purpose workstation was no longer sufficient: "Each activity setting supports a limited range of activities rather than all the things a person does" (Stone and Luchetti 1985, p. 106).

Today we know the concept of 'activity settings' as Activity Based Working (Skogland 2017). It has given rise to a diverse range of work settings that fulfil privacy and communication needs. As well as dedicated offices or hot desks, settings such as touchdown areas (short-stay non-bookable space), touchdown benches (high-level shared work areas), quiet rooms/booths, informal breakout spaces, restaurant areas, project spaces and meeting rooms are provided.

Instead of cashing in on the space savings made possible from hot-desking, many organisations have used the liberated spaces to support Activity Based Working. But this is not the end of the story. In order to be successful, Activity Based Working needs to stimulate a transformative process for individuals and organisations. It is no longer just about providing space. Designers and managers are challenged to make space 'readable'. Employees must make the right choices about where and how to work. Simply providing workplace choices without guidance makes the concept of Activity Based Working (ABW) 'window dressing'. Office workers will adopt routines that do not make use of the appropriate settings, choosing instead to resort to old or pre-existing habits. In order to break habits, an understanding of emotions is required.

In a study in the Netherlands (Hoendervanger et al. 2016), the researchers asked two key questions about Activity Based Working (ABW): Is switching behaviour related to satisfaction with ABW environments? Which factors explain switching behaviour? Their results indicated that "workers typically do not switch frequently, or not at all, between different activity settings … we observed a positive relationship between switching frequency and satisfaction with the ABW environment" (Hoendervanger et al. 2016, p. 49). They found that those employees that switched frequently derived greater satisfaction from the ABW environment. They also found that those employees with a diverse activity profile were more inclined to switch, as were people that were heavily dependent on communications. External mobility was the third job characteristic that increased switching frequency. Personality traits and temperaments were not examined in the study, with a focus on the more measurable job characteristics.

> We were told to do 'Activity Based Working' where teams would merge and move to a space depending on the project they were working on. But we found that a lot of those spaces are dead spaces not being used – as people gravitate to staying in one location. They feel they have a place in the organisation and by seeing the same people around them every time they have a sense of belonging.
> *Interview with public-sector officer reflecting on switching behaviour*

Can ABW environments improve our health? A recent study on the health effects of ABW (Engelen et al. 2019) concludes that "for physical and mental health, the evidence is equivocal … high-quality research is needed to strengthen the evidence base further and establish its health effects" (2019, p. 468). Unless switching becomes a frequent activity, the health benefits of moving around are not realised. Simply providing a diverse work setting is not enough.

The truly territorial office

If we read any literature about shared workspaces, we also see the term 'non-territorial office'. Yet this phrase misrepresents the modern work environment. It seems as though conflict over territory is more prominent than ever. Who has taken my files? Who is using the boardroom that I reserved? Why is he encroaching on my space? Why is she raising her voice on the phone and continually distracting me? The need and desire to become attached to territories is a fundamental part of the human psyche (Dittmar 1992). It comes as no surprise that we form strong attachment to things in the workplace, given how much time is spent there. We form attachments to work that we create, physical assets, spaces and job roles. We also go to considerable effort to establish territorial boundaries. Activities like defending our claims to things, conveying this to others and resolving conflicts form part of this 'effortful' landscape. Only when perceived territories are transgressed do we become consciously aware of these territories. We all see things differently – our perception of ownership, feelings of territory and the boundaries of our territories. But we also recognise as individuals the value of territories. From an instrumental perspective, they determine who gets to control a resource.

Emotion and the instrumental workspace 67

One emotion stands out in relation to territory in the workspace: anger. Anger plays a central role in influencing our experience of infringement. A key study by Brown and Robinson (2010) looked specifically at the emotion of anger linked to territory. It built on previous research suggesting that anger arises in situations of threat or thwarted goals, and prompts behaviours that seek to re-establish one's sense of self and place (Kitayama *et al.* 2000). Using the cognitive appraisal theory (Kuppens *et al.* 2003), they examined how people make sense of infringement (cognitive appraisals) and how this results in the emotional response of anger and the ensuing behaviours.

In the study, a number of 'reactionary defence behaviours' were observed including:

- using facial expressions to express disagreement or dislike towards the infringer;
- explaining to the infringer that the 'territory' was already claimed;
- involving co-workers to help reclaim the 'territory'.

The study revealed the diversity of territories that employees felt ownership over. This became evident when looking at infringement incidents or events. The most common form of territory that emerged as a result of infringement was physical objects (41 per cent) followed by work products and projects (14 per cent). Three important factors were found to determine the level of anger a person feels about encroachment in the workplace (as illustrated in Figure 6.2):

1 The extent to which one perceives an event is thwarting one's goals.
2 The assessment that an infringement event has been caused by some other person (as opposed to oneself or the situation).
3 The extent to which the infringement is seen as unfair or illegitimate.

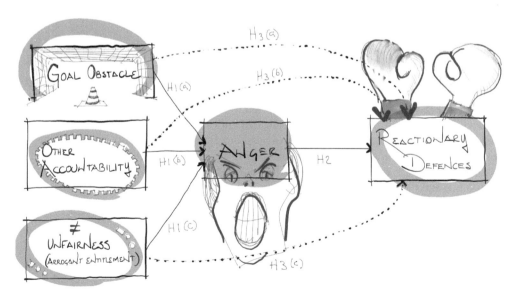

Figure 6.2 Determinants of anger in the territorial workplace.
Source: Based on Brown and Robinson 2010.

Looking at the first factor, instrumental objects (including space) are always perceived as being important to us because they are often essential for us to accomplish tasks. Furthermore, they may determine how we succeed in an organisation. Objects in the workplace can also have psychological importance, fulfilling a need for a place of our own, a sense of belonging and a means of self-expression. Psychological attachment and a sense of longing is pertinent to our symbolic understanding of emotions, as discussed in a later chapter. It embraces many 'positive emotions' rather than the 'negative emotions' of the instrumental dimension.

Considering the second factor, people inevitably ask, "Who is to blame?" when infringements occur. People are much more inclined to get angry if they feel the event was caused by another person. If a person is perceived to have infringed on a poorly marked territory, then they are less likely to be seen as responsible for an infringement. As designers we might consider how open plan designs can be used to clarify territorial boundaries. The use of task lighting, coloured carpet tiles and circulation lighting are examples of design interventions that can mark boundaries in open spaces. When infringers are seen as intentionally using or taking ownership of territory, anger is more likely to ensue.

The third factor that is likely to create anger is the perception that the infringement is unfair. This revolves around the idea of 'arrogant entitlement'. In such situations, other people are seen to take liberties – e.g. encroaching on space without permission. If, for example, a boss intrudes on another person's space, they may be exhibiting an unwelcome level of ownership prompting anger.

Clearly the myth of the 'non-territorial' office needs to be confronted. Shared space will inevitably give rise to emotions arising from the ambiguity of ownership. The cognitive appraisal framework provides a useful prompt for workplace designers. Understanding task requirements, perceptions of ownership, accountability and fairness all form part of the new melting pot that is the shared workplace. Designing shared environments places greater onus on the designer to reduce the ambiguity of space.

Behaviour change

The instrumental workplace has been transformed by new technology. But the possibilities that arise from innovations like Activity Based Working (ABW) rely on human habits and emotions. For example, what is the purpose of having many different work settings if individuals become entrenched in one preferred location? The appeal of the flexible workspace only remains in glossy interior design magazines or in the perception of occasional visitors. Can we provide clues to office users about how the space can be used? We can provide training and instigate online initiatives to promote new ways of working. But the evidence of success using these approaches is often muted. An initial enthusiasm and acceptance is typically followed by a return to old habits.

But it is often possible to provide clues about how an object should be used through the object itself or its context. This idea is referred to as 'affordance' (Gibson 2014). It does not rely on instruction manuals – simply encountering the artefact creates the desired behaviour. A more recent concept known as 'nudge' makes use of similar psychological ideas involving behaviour change. A number of examples illustrate how the technique of nudge can be used to change behaviour at work. But first, we might consider what the technique is and how it makes use of human decision-making (including affect).

Nudge

Choice is good – you can never have too much of it. That is the mantra that has been applied to office design and in the world at large. But we need to re-examine this assumption. We know, for example, that if we give people a choice about when and where to work, they will often choose the most inappropriate setting. Similarly, evidence suggests that people resist switching. In other words, people become comfortable with one arrangement, avoiding any perceived upheaval involved in changing. The concept of 'nudge' tackles just such a situation. It is an idea that emerged in the field of behavioural economics (Thaler and Sunstein 2008) and has given rise to several 'nudge' units at government level throughout the world. A nudge represents:

> Any aspect of the choice architecture that alters people's behaviour in a predictable way without forbidding any options or significantly changing their economic incentives. To count as a mere nudge, the intervention must be easy and cheap to avoid. Nudges are not mandates. Putting the fruit at eye level counts as a nudge. Banning junk food does not.
>
> (Thaler and Sunstein 2008, p. 6)

Choice architecture refers to the decision-making environment presented to the 'user'. Choice architects are the people responsible for 'framing' the decision-making environment. Whilst this may include traditional 'architects', choice architects refer to any person responsible for 'framing' the choice context. It can include those providing advisory information in public policy (e.g. pensions), the health sector, sustainability guidance, and health and safety, amongst others. One frequently cited example of nudge is the use of an engraved fly in the men's urinal at Schiphol airport. This intervention invariably improved the directional skills of male users, thus improving sanitary conditions. In general, the choice architect alters the environment in such a way that the intended behaviour is acted upon almost automatically.

Many organisations are investing in the sit–stand desk (SSD) in order to address health concerns about people in the office spending too much time in a fixed seated position. During the working day a typical European will spend 9½ hours sat down. Sedentary behaviour can lead to obesity, cardiovascular diseases and Type 2 diabetes, which together represent the leading causes of death and healthcare costs in Western societies (Wilmot et al. 2012). We might think that spending 30 minutes at the gym after work will compensate for our static existence in the office. Think again – research has shown that this approach may not overcome the unhealthy effects that arise from a sedentary lifestyle (Hamilton et al. 2008). Even those that consider themselves fit (e.g. those that cycle to work) may be subject to the risks associated with sitting for long periods.

A study by Wilk (1999) involving four different companies suggested that 60 per cent of their employees only used their SSD at standing height once a month or less – the main reason given being "they just did not bother to use the function". To investigate this assertion, a group in the Netherlands (Venema et al. 2018) looked at the use of SSDs. The researchers noted that the SSDs were always at sitting height. During the two-week period of their study, they changed the default sitting height to standing height, thereby introducing a 'default' nudge. A default nudge is effective because people do not have to expend effort in order to deal with the artefact. As the authors noted, "it makes use of people's inertia, while allowing the employee to place the desk back at sitting height if so desired"

(Venema et al. 2018, p. 671). As a result of their default nudge, stand-up working rates rose from 1.82 per cent to 13.13 per cent during the intervention. More importantly, a sustained level of 7.78 per cent was in evidence after two months.

Other examples where nudge has been used to address office environments include:

- social stairs: used the idea of an intelligent musical staircase to alter people's behaviour so that they used the stairs rather than the elevator – not only was this a health trigger, it also encouraged social engagement (Peeters et al. 2013);
- an engineering office: implemented a system known as EAST (easy, attractive, social, timely) in order to improve office-wide communication and increase job satisfaction (Healey et al. 2016);
- heating controls: examined the mental models of individuals when using heating control panels (Revell and Stanton 2014).

Summary

When we think of the emotions associated with getting work done, we think of frustration sometimes leading to anger. As designers, all we can do is minimise dissatisfaction and make sure that the tools and environment enable users to get their work done. This hygiene perspective persists in many service level agreements (SLAs) and is often embodied in the work environments we create. When we measure satisfaction, we are in fact measuring levels of dissatisfaction in the hope that we can eliminate problems. However, problems persist – even in supposedly cutting-edge workplaces – that cannot be resolved with a purely instrumental perspective. The relationship between technology and humanity sometimes appears to be broken. In the remainder of this book we look at emotions more keenly associated with 'positive psychology'.

References

Allen, T., 2004. *Working Without Walls: An Insight into Transforming Government Workplace*. London: Office of Government Commerce.

Brown, G., 2009. Claiming a corner at work: Measuring employee territoriality in their workspaces. *Journal of Environmental Psychology*, 29 (1), 44–52.

Brown, G. and Robinson, S.L., 2010. Reactions to territorial infringement. *Organization Science*, 22 (1), 210–224.

Demerouti, E. and Bakker, A.B., 2007. The job demands–resources model: State of the art. *Journal of Managerial Psychology*, 22 (3), 309–328.

Dittmar, H., 1992. *The Social Psychology of Material Possessions: To Have Is to Be*. Hemel Hempstead: Harvester Wheatsheaf and St. Martin's Press.

Does 5S make sense in an office environment?, 2012. *The Workplace Stack Exchange* [online]. Available from: https://workplace.stackexchange.com/questions/431/does-5s-make-sense-in-an-office-environment [accessed 10 December 2018].

Eason, K.D., 1995. User-centred design: For users or by users? *Ergonomics*, 38 (8), 1667–1673.

Elsbach, K.D., 2003. Relating physical environment to self-categorizations: Identity threat and affirmation in a non-territorial office space. *Administrative Science Quarterly*, 48 (4), 622–654.

Engelen, L., Chau, J., Young, S., Mackey, M., Jeyapalan, D. and Bauman, A., 2019. Is activity-based working impacting health, work performance and perceptions? A systematic review. *Building Research & Information*, 47 (4), 468–479.

Gibson, J.J., 2014. *The Ecological Approach to Visual Perception: Classic Edition*. London: Psychology Press.

Hamilton, M.T., Healy, G.N., Dunstan, D.W., Zderic, T.W. and Owen, N., 2008. Too little exercise and too much sitting: Inactivity physiology and the need for new recommendations on sedentary behavior. *Current Cardiovascular Risk Reports*, 2 (4), 292.

Hardy, B., 2008. *Working Beyond Walls: The Government Workplace as an Agent of Change*. London: Office of Government Commerce; DEGW.

Hay, R., Samuel, F., Watson, K.J. and Bradbury, S., 2018. Post-occupancy evaluation in architecture: Experiences and perspectives from UK practice. *Building Research & Information*, 46 (6), 698–710.

Healey, M., Parker, K. and Saldin, L., 2016. Using nudge techniques to influence behaviour in an engineering office. *Society of Petroleum Engineers (SPE) Intelligent Energy International Conference and Exhibition*, Aberdeen.

Hirst, A., 2011. Settlers, vagrants and mutual indifference: Unintended consequences of hot-desking. *Journal of Organizational Change Management*, 24 (6), 767–788.

Hoendervanger, J.G., De Been, I., Van Yperen, N.W., Mobach, M.P. and Albers, C.J., 2016. Flexibility in use: Switching behaviour and satisfaction in activity-based work environments. *Journal of Corporate Real Estate*, 18 (1), 48–62.

Jackson, S.L., 2010. *The Men: American Enlisted Submariners in World War II. Why They Joined, Why They Fought, and Why They Won*. Indianapolis, IN: Dog Ear Publishing.

Kitayama, S., Markus, H.R. and Kurokawa, M., 2000. Culture, emotion and well-being: Good feelings in Japan and the United States. *Cognition and Emotion*, 14 (1), 93–124.

Kuppens, P., Van Mechelen, I., Smits, D.J.M. and De Boeck, P., 2003. The appraisal basis of anger: Specificity, necessity and sufficiency of components. *Emotion*, 3 (3), 254–269.

Morrison, R.L. and Macky, K.A., 2017. The demands and resources arising from shared office spaces. *Applied Ergonomics*, 60, 103–115.

Peeters, M., Megens, C., van den Hoven, E., Hummels, C. and Brombacher, A., 2013. Social stairs: Taking the piano staircase towards long-term behavioral change. In: S. Berkovsky and J. Freyne, eds, *Persuasive Technology*. Springer Berlin Heidelberg, 174–179.

Preiser, W.F.E., White, E. and Rabinowitz, H., 2015. *Post-Occupancy Evaluation (Routledge Revivals)*. London and New York: Routledge.

Revell, K.M.A. and Stanton, N.A., 2014. Case studies of mental models in home heat control: Searching for feedback, valve, timer and switch theories. *Applied Ergonomics*, 45 (3), 363–378.

Sanders, E.B.-N. and Stappers, P.J., 2008. Co-creation and the new landscapes of design. *CoDesign*, 4 (1), 5–18.

Skogland, M.A.C., 2017. The mindset of activity-based working. *Journal of Facilities Management*, 15 (1), 62–75.

Stone, P.J. and Luchetti, R., 1985. Your office is where you are. *Harvard Business Review*, 63 (2), 102–117.

Thaler, R.H. and Sunstein, C.R., 2008. *Nudge: Improving Decisions about Health, Wealth and Happiness*. New Haven, CT: Yale University Press.

Venema, T.A.G., Kroese, F.M. and De Ridder, D.T.D., 2018. I'm still standing: A longitudinal study on the effect of a default nudge. *Psychology & Health*, 33 (5), 669–681.

Wilk, J., 1999. Mind, nature and the emerging science of change: An introduction to metamorphology. In: G.C. Cornelis, S. Smets and J.P. Van Bendegem, eds, *Metadebates on Science: The Blue Book of 'Einstein Meets Magritte'*. Dordrecht: Springer Netherlands, 71–87.

Wilmot, E.G., Edwardson, C.L., Achana, F.A., Davies, M.J., Gorely, T., Gray, L.J., Khunti, K., Yates, T. and Biddle, S.J.H., 2012. Sedentary time in adults and the association with diabetes, cardiovascular disease and death: Systematic review and meta-analysis. *Diabetologia*, 55 (11), 2895–2905.

7 Emotion and the aesthetic workspace

The aesthetic features of a workspace are often seen as a desirable but not essential part of any working environment. It is often intermingled with ideas of luxury and indulgence which perhaps distract from the business of work. In this chapter we examine the research evidence showing how the 'look and feel' of the workplace affects our everyday well-being and decision-making. We illustrate how environmental aesthetics can spark creativity. Unlike the future-oriented instrumental workplace, the aesthetic workplace provides the sensory stimulus in order to function in the present.

Being there

The previous chapter was wholly concerned with 'doing' but this chapter is about 'being'. A sceptic might say the only reason a person occupies a workspace is to be productive. They might argue that the workspace's sole purpose is to provide the necessary tools and setting to get the work done, without distraction.

Let's look at some illuminating facts about the modern-day office worker. The average American spends 90,000 hours of their lifetime at work (Pryce-Jones 2011). In Japan, hundreds of Japanese office workers die every year from *karoshi*, otherwise known as death by overwork. And in the UK, a third of managers are losing their sense of humour because of work (Worrall et al. 2016).

Many commentators have expressed concern that people in the world of work have become disconnected. In the book entitled *The Corrosion of Character*, Sennett (2000) asks what happens in a society where the main characteristics of economic activity and work are neither long-term nor stable, but flexible and short-term? He argues that the emphasis on short-term projects and contract performance is reformulating people's identities and their integrity with regard to changes in working life. In an article appearing in the *Harvard Business Review* (Coleman and Coleman 2016), the problem of work stress becoming home stress was flagged up. We like to think we can compartmentalise work stress and contain emotions. But emotional expressions are often not tolerated at work. As a result, much of the stress from work ends up being taken out on partners, children and friends. In a 'click-out and forget' world, Courpasson (2016, p. 1095) observes that "we tend to physically disappear behind our screens while expressing these emotions in our curious isolation; we are contained and confined by our screens, in the secluded offices where we spend most of our time".

What do we really mean by being 'present'? The phenomena of 'presenteeism' has become widespread in organisations. It refers to being present at one's place of work for more hours than is required, especially as a result of insecurity about our job. We are there

Emotion and the aesthetic workspace 73

– and at the same time we are not there! We may be present in body but not in mind. We concoct ways of switching off from the world around us – primarily to avoid distractions in the hubbub of the workplace. Yet as nonroutine creative jobs supplant routine work, we increasingly need to invest our heart and mind.

Take for example the practice of mutual indifference in the open plan office – we have all done it. It forms part of modern-day office etiquette. Otherwise known as 'civil inattention' (Goffman 1971), mutual indifference is what we often practice on public transport. We look away; we do not acknowledge our nearest neighbour. The practice is reciprocated and we both understand we have work to do. But at the same time, our withdrawal results in a withdrawal of commitment to the organisation. Hirst observes some irony in this:

> ...People showing civil inattention are actively avoiding networking. ... This issue is potentially significant because civil inattention contrasts so markedly with the intensely networked communication which it is claimed is needed in contemporary organisations, and which is claimed to result from the architectural design of the new office.
>
> (Hirst 2011, p. 784)

What does it take to engage people? It seems that people's attempts to avoid distraction are undermining their presence. There is a single word in the German vocabulary known as *Dasein* that simply means 'existence', as in 'I am pleased with my existence'. Another interpretation for the word *Dasein* is 'being there'. The German philosopher Martin Heidegger thought of *Dasein* as a way of "being involved with and caring for the immediate world in which one lives" (Collins *et al.* 1998). This 'authentic' view of the world involved one's individuality, one's own limited lifespan, and one's own being. It contrasted with the everyday and inauthentic way of living which involves giving up one's own individual identity in favour of an 'escapist' immersion in the public everyday world – the anonymous, identical world of the 'they and them'.

I have noticed amongst the younger workers that they have all got their headphones on now, headphones and iPods. They put them on to disengage from background conversation. It is a bit like ... "if you have placed us in this open plan where we now have lost our capacity to concentrate, we are now going to blank-out the distractions". It does not help people to focus or collaborate more. I think that actually has added a whole bunch of stimuli which is unnecessary for the work we do.
Interview with public-sector officer working in shared space

From an instrumental perspective, we think of the workplace simply as equipment providing a particular system of meaning and purpose (affordance). For example, when an office chair is comfortable and supportive, we cease to be aware of it. Heidegger refers to this as 'ready to hand', providing an oversimplified presence when reduced to possible future usefulness to us. We can also start to think of the people around us whose presence is only noted by their possible usefulness to us. Our emotions arise through a disruptive process mediated by judgement (cognition) when things go wrong.

Aesthetics and the senses

The aesthetic dimension invites more positive emotions. It allows us to live 'in the moment' by the direct impact of the senses without intervention by thought processes. Seen from this perspective, the workplace aesthetic is not concerned with ornamentation or luxury but as a necessary part of any creative workplace founded on 'being there'. It provides us with a context that puts us in contact with ourselves.

We invariably think of sight and sound when making aesthetic emotional judgements (like joyful, pleasant, arousing, boring or annoying). In reality there are five sense modalities that exist – Visual, Acoustic, Kinaesthetic, Olfactory and Gustatory – and a number of sub-modalities (Nielsen and Nielsen 2011):

V Colour, brightness, size, location, shape, intensity.
A Sound/melody, volume, theme, word, sentence.
K Sensation, respiration, temperature, motion/rest, weight, gesture.
O Fresh, fruity, flowery, perfumed.
G Spicy, mild, sweet, tasty.

Many studies have sought to understand how different environmental factors can be used to impart a positive emotional state at work. Environmental characteristics have attracted attention, such as: natural lighting and artificial lighting (Aries et al. 2015, Bille 2015, Yang and Fotios 2015, Figueiro and Rea 2016); colour (Bakker et al. 2015); sound and music (Lesiuk 2005, Haake 2011, Kang et al. 2016); and aroma and fragrance (Baron and Bronfen 1994, Knasko 1997, Lehrner et al. 2000, Riach and Warren 2015).

But if we want to provide sensory stimulation in order to achieve a desired behaviour or experience, things are not that simple. Most research on human emotional responses are carried out in highly controlled conditions and do not reflect the complexity of real-life environments. People experience their surroundings in a dynamic and interactive way (Schreuder et al. 2016). Moreover, all environments provide a multisensory stimulus. The sensory input that we experience is not simply perceived as the sum of its individual sensory modes. There appears to be a broad spectrum of non-linear interactions that occur between each of the sensory modalities. Interactions between colour, shades, odours and noise combine so that effects are multiplied (sensory cooperation), disambiguated (one sensory stimulant helps make sense of another stimulant), vetoed (a stronger stimulant overrides a weak one) or novel effects occur (as shown in the McGurk effect whereby hearing one sound with a visual component gives rise to a perceived third sound (McGurk and Macdonald 1976)).

Aesthetics and emotions

We know that an aesthetic emotion is something that is felt rather than known. It relies on the immediacy of the senses. Aesthetic emotions play a part in music, literature, film, painting and built environments. But it seems no agreement exists regarding a universal definition. Instead, we rely on exemplars or prototypes for understanding aesthetic appeal. We can identify universal feelings such as fascination, awe and beauty, and negative emotions such as confusion, ugliness and boredom. Whilst the categorisation of aesthetic emotions is often disparate, there are several features that distinguish aesthetic emotions from the utilitarian emotions discussed in the previous chapter. These include (Schindler et al. 2017):

- An aesthetic emotion is something that we feel rather than something that is represented, expressed or alluded to.
- In contrast to instrumental emotions, aesthetic emotions arise from an object's intrinsic aesthetic appeal. It does not depend on our personal or organisational goals.
- Aesthetic emotions are more heavily influenced by 'distance senses' (i.e. visual and auditory) as opposed to 'contact senses' (touch, taste and smell).
- Whilst aesthetic emotions primarily arise 'in the moment', there is also a process of judgement or 'art framing' that often goes hand-in-hand. However, this top-down judgement process is not always a necessary part of aesthetic emotions.

We also need to consider how we interact with our workspace. In contrast with the static appreciation of art in a gallery, aesthetic emotions (positive emotions) often arise from dynamic interactions (Desmet 2012):

- positive emotions evoked by the material qualities of the surrounding work environment;
- positive emotions evoked by personal meanings associated with the environment (interrelated with symbolic dimension discussed in the next chapter);
- positive emotions evoked by interactive qualities of using the workspace;
- positive emotions evoked by activities enabled or facilitated by the space;
- positive emotions evoked by ourselves as users of the space;
- positive emotions evoked by the effects of other people's activities on us, in which the environment plays some role.

In a complex interactive environment like the workspace, it is often difficult to draw a clear distinction between aesthetic emotions, instrumental emotions and symbolic emotions. It seems that the perceived attractiveness of products also enhances their usability. In terms of brain chemistry, emotions triggered by objects rather than outcomes stimulate an entirely different neural system. The aesthetic emotion stimulated by objects prompt activity in the 'liking system', whilst the instrumental emotions are prompted by the 'wanting system' (Berridge and Kringelbach 2008).

Biophilia and biomorphic design

Have you ever wondered why it is so difficult to concentrate at work? Rather than 'being there', you feel like you are 'elsewhere'. Many modern workplaces can be described as inorganic environments: ones that show an "excessive reliance on fabricated materials, artificial lighting, controlled climatic conditions, straight-line geometries, homogeneity of design, scales rarely if ever encountered in nature, and substitution of the synthetic or the natural" (Kellert 2003).

A growing number of researchers now believe that a work environment that embraces the aesthetic qualities of nature may be the answer. The psychological advantages of immersion in natural environments appear to go well beyond simply 'fixing' the consequences of work overload (stress). According to Appleton's habitat theory (Appleton 1996) our psychological responses to architectural spaces have evolved from our ancestors who lived in savannah-type environments. These landscapes possessed visual features and spatial configurations that favoured survival. This goes some way to explaining why humans prefer settings that have some degree of complexity, but can also be easily or

76 Emotion and the aesthetic workspace

fluently processed. We appear to have an innate sensitivity to features like contrast, grouping and symmetry (see Figure 7.1).

Can natural workplace environments be more than just a fad? A growing body of evidence suggests that organic workplaces can: improve mood (Aerts *et al.* 2018); reduce stress (Valtchanov *et al.* 2010); increase feelings of vitality and energy (Ryan *et al.* 2010); improve concentration and working memory (Ayuso Sanchez *et al.* 2018); as well as improve people's self-esteem (Readdick and Schaller 2005). Many of the benefits address the 'flourishing' or self-actualising properties of the aesthetic dimension.

But that does not mean just sticking a pot plant by the photocopier. The interest in 'organic' environments includes biophilic and biomorphic design. Some environments embrace nature-like characteristics based on the use of naturalistic forms and patterns. In a biomorphic design, the intention is to apply aesthetic mimicry rather than inclusion of natural artefacts. This biomorphic design approach creates visual patterns that are distributed throughout an architectural setting. In contrast, a biophilic design makes use of discrete elements such as water features, vegetation and sunlight – or even pets.

> For me, it's all about having natural light and being able to access quiet spaces to work. Being able to see things like plants or the natural environment is very important to me. When I moved into an office, I had an internal office that did not have that and I very much struggled in that environment.
> *Interview with worker in start-up space following 20 years' experience in legal corporate environments*

Figure 7.1 Creating 'naturalness' in the work environment.

Emotion and the aesthetic workspace

Two key theories explore the aesthetic value of 'naturalness'. The first of these is based on the Biophilia Hypothesis (BH) (Kellert and Wilson 1995). The word biophilia means 'love of life', expressing the clear connection between human emotions and nature. The hypothesis contends that we are genetically programmed to want to be close to plants, animals and natural surroundings. The related theory known as Attention Restoration Theory (ART) considers the effects that arise from interacting with nature. It considers how our limited cognitive resources can be replenished through natural stimuli. In this way we are able to restore our attentional resources and perform better at our job (Kaplan 1995).

If we are able to measure 'naturalness', then designers of workspaces should be able to integrate the sensory benefits of such spaces. But the task of measuring the degree of 'naturalness' and the resulting emotional impact is challenging – until recently it has defied objective measurement. Both psychological measurements and measurements based on neuroscience (neuroaesthetics and neuroarchitecture) promise to fill this knowledge gap.

Fractals

It seems that nature uses a hidden code. A property known as 'fractal geometry' is found in abundance in natural environments. Fractals refer to "fractured shapes that possess repeating patterns when viewed at increasingly fine magnifications" (Hagerhall *et al.* 2004, p. 247). As children we might have discovered how to produce elaborate patterns based on a simple set of cogs and wheels using a Spirograph. Natural systems create similarly complex imagery at varying scales. It is possible to use statistical measures to analyse the fractal dimensions of visual elements. For example, a simple curve has a fractal dimension of approximately 1, whereas a densely convoluted line has a fractal dimension of around 2. This provides a measure of complexity in our surroundings. This complexity enables us to identify objects and extract information from the built environment. If we strip out these

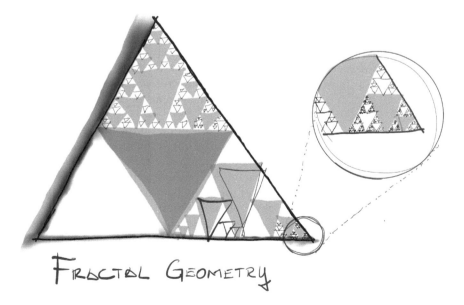

Figure 7.2 Fractal geometry.

cues and create workplace environments without visual complexity, it seems that we deprive our visual systems access to meaningful information. In many 'hygienic' workspaces, the reliance on Euclidean geometry, straight lines and uniformity banishes visual complexity. Too much architectural complexity also appears to overwhelm our visual systems, suggesting that there is a Goldilocks region involving a moderate level of visual complexity (Berlyne and Boudewijns 1971). When it comes to measuring aesthetic 'beauty', it seems that complexity occurs at three different levels: (1) the amount and variety of elements; (2) the way those elements are organised; and (3) the level of asymmetry (Nadal et al. 2010).

A recent study sought to measure the 'naturalness' of environments using diffuse spatial and colour features in architectural scenes (Coburn et al. 2019). Could subjective perceptions of naturalness in architectural scenes be measured objectively? The intention in the study was to ascertain the underlying aesthetic dimensions that influence how we respond to a scene. Participants were asked to undertake an image arrangement task for natural and man-made scenes. The scenes were analysed for: (1) scaling; and (2) contrast-related features including fractal density. It was found that these two features were strong predictors of naturalness for both interior and exterior scenes. Naturalness ratings appear to explain more than half of the variation in ratings, suggesting that people invoke an innate perception of naturalness. Moreover, the study highlights how image analysis can be used to objectively measure the aesthetic appeal of a building's interior.

Neuroaesthetics and neuroarchitecture

They say a picture is worth a thousand words. But sometimes a picture is not enough when it comes to understanding emotions – particularly in architectural settings. Studies using photographs to portray different scenescapes provide only a static 2D impression for the participants. Respondents in such studies are only responding to a flat portrayal of real-life. But immersive technologies enable us to overcome these constraints.

> Each different medium with which we analyse and map spaces offers a different insight, and can potentially increase our tools and methods for mapping spaces and understanding human experience. The emergence of such technologies has the potential to influence the way in which we map, analyse and perceive spaces.
> (Karandinou and Turner 2017, p. 54)

Developments in neuroscience have delivered a new window on the brain, with a tantalising clue about how the mind works. The development of functional MRI (fMRI) promises to offer up new insights. fMRI is based on the same technology as magnetic resonance imaging (MRI) encountered in hospitals. Such scanners emit a strong magnetic field and radio waves to create detailed images of the body. But instead of creating images of organs and tissues, as MRI does, fMRI looks at blood flow in the working brain to detect areas of activity. Another technology that is used to map the brain is EEG (electroencephalography). EEG uses electrodes that are placed around the scalp. A drawback for EEG is the spatial resolution – as the electrodes measure electrical activity at the surface of the brain, it is difficult to know whether the signal was produced near the surface (in the cortex) or from a deeper region of the brain.

These two imaging technologies have given rise to new disciplines in the twenty-first century, including the nascent sciences of neuroaesthetics and neuroarchitecture. The advent of neuroaesthetics has allowed researchers to evaluate the neural responses to art. Using fMRI, researchers have for the first time been able to undertake empirical as well as theoretical studies about human responses to artwork (Kirk et al. 2009). But built environments present a markedly different aesthetic experience (and thus a distinct measurement challenge):

- The experience of architectural space is immersive. We do not just stand back and observe – we inhabit the space.
- It is multisensory in nature, involving several senses in addition to the visual stimulus.
- It involves interaction over a prolonged lifespan (compared with short encounters with artwork). Our aesthetic and emotional responses invariably evolve in 'lived-in' space.

The unique challenges of man-made environments have necessitated the development of a new science known as neuroarchitecture (Eberhard 2009). A number of significant studies have emerged in this discipline:

- Wayfinding using a portable EEG (electroencephalography). The study examined how ten participants undertook four different journeys. It examined key decision-making moments in their journey and how this related to recordings of specific brain waves (Karandinou and Turner 2017).
- A study examining the impact of different interior forms on people's affective state and the associated brain activity. EEG was used in conjunction with virtual reality (VR) to examine how people explore and experience architectural spaces. The results suggested a strong impact of curvature geometries on the activity of the anterior cingulate cortex (ACC) (Banaei et al. 2017).
- An investigation into people's fear response in a healthcare environment. The results suggest significantly higher amygdala activation associated with sharp contours versus curved contours. fMRI was used in this study, providing a higher level of brain image resolution (Pati et al. 2016).
- A novel study examining how different architectural styles impact on the brain. The findings demonstrate how the human visual system encodes visual aspects of architecture (Choo et al. 2017).

A word of warning when using neuroscientific evidence! It has become a highly popularised discipline moving into many mainstream areas including neuromarketing, neuroeconomics and neuroarchitecture. Brain imaging using colour-drenched images can give an impression of scientific rigour to any research. Three experiments undertaken by McCabe and Castel (2008) suggested that simply displaying brain images in articles resulted in higher ratings of scientific reasoning for arguments made in those articles. Popular science often suggests that we can now decode the brain – that we can see with our own eyes emotions such as love or delight or mistrust. This reductionist explanation is very appealing to the layman. The reality is that we cannot look in the brain to see what it is doing. The image of the brain is not the image of the mind. All we are seeing are the part of the brain where the oxygen levels have changed the most. The popularisation of neuroscience has led to overpromising (van Atteveldt et al. 2014). But used with caution, neuroscience is providing new insights about environment–behaviour interaction – often supporting established evidence from the world of psychology and linguistics.

Summary

Interaction with our workspace takes many forms. We engage several sensory modes that combine to produce an emotional response. Whilst a significant amount of work has been done to assess how specific environmental factors affect our experience of space, the combined effect of different senses is less understood. However, a growing body of evidence demonstrates how positive emotions can be evoked by interaction with the physical environment. This aesthetic experience has previously been seen as ornamental and marginal to the everyday practice of business. More and more we see that our experiential or corporeal presence is paramount. The office of the future needs to capture the office of the present.

References

Aerts, R., Honnay, O. and Van Nieuwenhuyse, A., 2018. Biodiversity and human health: Mechanisms and evidence of the positive health effects of diversity in nature and green spaces. *British Medical Bulletin*, 127 (1), 5–22.

Appleton, J., 1996. *The Experience of Landscape*. Chichester: John Wiley & Sons.

Aries, M., Aarts, M. and van Hoof, J., 2015. Daylight and health: A review of the evidence and consequences for the built environment. *Lighting Research & Technology*, 47 (1), 6–27.

Atteveldt, N.M. van, van Aalderen-Smeets, S.I., Jacobi, C. and Ruigrok, N., 2014. Media reporting of neuroscience depends on timing, topic and newspaper type. *PLOS ONE*, 9 (8), e104780.

Ayuso Sanchez, J., Ikaga, T. and Vega Sanchez, S., 2018. Quantitative improvement in workplace performance through biophilic design: A pilot experiment case study. *Energy and Buildings*, 177, 316–328.

Bakker, I., van der Voordt, T., Vink, P., de Boon, J. and Bazley, C., 2015. Color preferences for different topics in connection to personal characteristics. *Color Research & Application*, 40 (1), 62–71.

Banaei, M., Hatami, J., Yazdanfar, A. and Gramann, K., 2017. Walking through architectural spaces: The impact of interior forms on human brain dynamics. *Frontiers in Human Neuroscience*, 11, n.p.

Baron, R.A. and Bronfen, M.I., 1994. A whiff of reality: Empirical evidence concerning the effects of pleasant fragrances on work-related behavior. *Journal of Applied Social Psychology*, 24 (13), 1179–1203.

Berlyne, D.E. and Boudewijns, W.J., 1971. Hedonic effects of uniformity in variety. *Canadian Journal of Psychology/Revue canadienne de psychologie*, 25 (3), 195–206.

Berridge, K.C. and Kringelbach, M.L., 2008. Affective neuroscience of pleasure: Reward in humans and animals. *Psychopharmacology*, 199 (3), 457–480.

Bille, M., 2015. Lighting up cosy atmospheres in Denmark. *Emotion, Space and Society*, 15, 56–63.

Choo, H., Nasar, J.L., Nikrahei, B. and Walther, D.B., 2017. Neural codes of seeing architectural styles. *Scientific Reports*, 7, 40201.

Coburn, A., Kardan, O., Kotabe, H., Steinberg, J., Hout, M.C., Robbins, A., Macdonald, J., Hayn-Leichsenring, G. and Berman, M.G., 2019. Psychological responses to natural patterns in architecture. *Journal of Environmental Psychology*, 62, 133–145.

Coleman, J. and Coleman, J., 2016. Don't take work stress home with you. *Harvard Business Review*, 28 July, 2–4.

Collins, J., Appignanesi, R. and Selina, H., 1998. *Heidegger for Beginners*. London: Icon Books.

Courpasson, D., 2016. Looking away? Civilized indifference and the carnal relationships of the contemporary workplace. *Journal of Management Studies*, 53 (6), 1094–1100.

Desmet, P.M.A., 2012. Faces of product pleasure: 25 positive emotions in human–product interactions. *International Journal of Design*, 6 (2), n.p.

Eberhard, J.P., 2009. *Brain Landscape: The Coexistence of Neuroscience and Architecture*. New York: Oxford University Press.

Figueiro, M. and Rea, M., 2016. Office lighting and personal light exposures in two seasons: Impact on sleep and mood. *Lighting Research & Technology*, 48 (3), 352–364.

Goffman, E., 1971. *Relations in Public: Micro-Studies of the Public Order*. London: Penguin.

Haake, A.B., 2011. Individual music listening in workplace settings: An exploratory survey of offices in the UK. *Musicae Scientiae*, 15 (1), 107–129.

Hagerhall, C.M., Purcell, T. and Taylor, R., 2004. Fractal dimension of landscape silhouette outlines as a predictor of landscape preference. *Journal of Environmental Psychology*, 24 (2), 247–255.

Hirst, A., 2011. Settlers, vagrants and mutual indifference: Unintended consequences of hot-desking. *Journal of Organizational Change Management*, 24 (6), 767–788.

Kang, J., Aletta, F., Gjestland, T.T., Brown, L.A., Botteldooren, D., Schulte-Fortkamp, B., Lercher, P., van Kamp, I., Genuit, K., Fiebig, A., Bento Coelho, J.L., Maffei, L. and Lavia, L., 2016. Ten questions on the soundscapes of the built environment. *Building and Environment*, 108, 284–294.

Kaplan, S., 1995. The restorative benefits of nature: Toward an integrative framework. *Journal of Environmental Psychology*, 15 (3), 169–182.

Karandinou, A. and Turner, L., 2017. Architecture and neuroscience: What can the EEG recording of brain activity reveal about a walk through everyday spaces? *International Journal of Parallel, Emergent and Distributed Systems*, 32 (sup1), S54–S65.

Kellert, S.R., 2003. *Kinship to Mastery: Biophilia In Human Evolution and Development*. Washington, DC: Island Press.

Kellert, S.R. and Wilson, E.O., 1995. *The Biophilia Hypothesis*. Washington, DC: Island Press.

Kirk, U., Skov, M., Hulme, O., Christensen, M.S. and Zeki, S., 2009. Modulation of aesthetic value by semantic context: An fMRI study. *NeuroImage*, 44 (3), 1125–1132.

Knasko, S.C., 1997. Ambient odour: Effects on human behaviour. *International Journal of Aromatherapy*, 8 (3), 28–33.

Lehrner, J., Eckersberger, C., Walla, P., Pötsch, G. and Deecke, L., 2000. Ambient odor of orange in a dental office reduces anxiety and improves mood in female patients. *Physiology & Behavior*, 71 (1), 83–86.

Lesiuk, T., 2005. The effect of music listening on work performance. *Psychology of Music*, 33 (2), 173–191.

McCabe, D.P. and Castel, A.D., 2008. Seeing is believing: The effect of brain images on judgments of scientific reasoning. *Cognition*, 107 (1), 343–352.

McGurk, H. and Macdonald, J., 1976. Hearing lips and seeing voices. *Nature*, 264 (5588), 746.

Nadal, M., Munar, E., Marty, G. and Cela-Conde, C.J., 2010. Visual complexity and beauty appreciation: Explaining the divergence of results. *Empirical Studies of the Arts*, 28 (2), 173–191.

Nielsen, K. and Nielsen, N., 2011. *NLP: Die Karten zur NLP – Ausbildung*. Berlin: Heragon.

Pati, D., O'Boyle, M., Hou, J., Nanda, U. and Ghamari, H., 2016. Can hospital form trigger fear response? *HERD: Health Environments Research & Design Journal*, 9 (3), 162–175.

Pryce-Jones, J., 2011. *Happiness at Work: Maximizing Your Psychological Capital for Success*. Chichester: John Wiley & Sons.

Readdick, C.A. and Schaller, G.R., 2005. Summer camp and self-esteem of school-age inner-city children. *Perceptual and Motor Skills*, 101 (1), 121–130.

Riach, K. and Warren, S., 2015. Smell organization: Bodies and corporeal porosity in office work. *Human Relations*, 68 (5), 789–809.

Ryan, R.M., Weinstein, N., Bernstein, J., Brown, K.W., Mistretta, L. and Gagné, M., 2010. Vitalizing effects of being outdoors and in nature. *Journal of Environmental Psychology*, 30 (2), 159–168.

Schindler, I., Hosoya, G., Menninghaus, W., Beermann, U., Wagner, V., Eid, M. and Scherer, K.R., 2017. Measuring aesthetic emotions: A review of the literature and a new assessment tool. *PLOS ONE*, 12 (6), e0178899.

Schreuder, E., van Erp, J., Toet, A. and Kallen, V.L., 2016. Emotional responses to multisensory environmental stimuli: A conceptual framework and literature review. *SAGE Open*, 6 (1), 2158244016630591.

Sennett, R. and Kovalainen, A., 2000. Book reviews: "The Corrosion of Character: The Personal Consequences of Work in the New Capitalism". *Acta Sociologica*, 43 (2), 175–177.

Valtchanov, D., Barton, K.R. and Ellard, C., 2010. Restorative effects of virtual nature settings. *Cyberpsychology, Behavior, and Social Networking*, 13 (5), 503–512.

Worrall, L., Cooper, C., Kerrin, M., La-Band, A., Rosseli, A. and Woodman, P., 2016. *The Quality of Working Life: Exploring Managers' Wellbeing, Motivation and Productivity*. London: Chartered Management Institute.

Yang, B. and Fotios, S., 2015. Lighting and recognition of emotion conveyed by facial expressions. *Lighting Research & Technology*, 47 (8), 964–975.

8 Emotion and the symbolic workspace

Spaces and objects in the workplace have the power to evoke memories and thoughts. Our initial sensation of an artefact triggers cognitive thought processes because it carries meaning. In this way we are able to make sense of our environment. This is what we call symbolism. We typically perceive the world around us through our own unique lens, both individually and as part of a culture. This chapter examines how the symbolic aspect of the workspace can be used to enable sensemaking in an increasingly disruptive world.

What is a symbol?

Edmund Spenser, an English poet, in 1590 wrote a very concise definition of what is meant by a symbol:

> Something which stands for something else.
>
> (Spenser 2007)

We use words, sounds, gestures and visual images as symbols. The workplace itself and the elements within it have a symbolic function. Symbols in our place of work provide linkages between seemingly unrelated concepts and experiences. These linkages allow us to go beyond what is known or seen. Unlike a sign which has only one meaning, a symbol often has multiple levels of meaning. A symbol can be described as a "visual image or sign representing an idea … a deeper indicator of a universal truth" (Bruce-Mitford 2008, p. 6).

Have you ever noticed how carpet tiles are used to help us navigate an open plan office? One colour might be used to indicate circulation paths, whilst other colours might be used to mark out boundaries or zones. We might also use localised lighting or circulation lighting to reinforce these boundaries so that even a new visitor is left in no doubt where they need to go. The symbolic wall replaces the physical wall in the open plan space. Whilst there is no tangible obstacle preventing us from straying across boundaries, we understand the symbolic significance of the carpet tiles.

> A symbol is an energy evoking and directing agent.
>
> (Campbell 2002, p. 143)

Unlike instrumentalism that is future oriented or aesthetics that is concerned with the present, symbolism relies on the past. We rely on an accumulation of previous experiences and form attachments to familiar objects and surroundings.

Symbols often allow complex communication at multiple levels of meaning. They allow organisations to convey ideologies and structures. They evoke emotions that are culturally learned, in contrast to the aesthetic experience that is innate and spontaneous. Rather than direct sensory input, there is an intertwining of sensory and cognitive processes that give rise to our emotional interpretation. We attach meaning and memories to surroundings. We accumulate experiences that produce an evolving symbolic association.

> I do have a little wooden sculpture that I bring in – which is a little man with a backpack and a walking stick – and that reminds me that I'm not tied to my desk and my job is not me … it's gone with me to every job I've had for the last 20 years.
> *Interview with consultant occupying start-up incubator space*

Symbolism and the organisation

Symbolism enables us to function collaboratively. It is a social mechanism that enables us to express ourselves. The idea was aptly expressed by the psychologist William James in 1890 who pursued the idea of the 'empirical self' – how people think about themselves. He suggested that there were three different categories: (1) the 'material self'; (2) the 'spiritual self'; and (3) the 'social self' (James 2013).

Making sense of clues in the workplace became a mainstream part of architectural studies with the publication of the book *Inquiry by Design* by John Zeisel (1984). This was pushed further by key publications on post-occupancy evaluation (including Preiser 1995) that demonstrated how a systematic appraisal could be used to evaluate building performance – incorporating symbolic as well as functional indicators. Not only was it a design tool, it was an activity to be scheduled by facilities managers as part of a building's evolution. The importance of physical space as a vehicle for communication is expressed emphatically by Gagliardi (2011, p. vi):

> Space and artefacts constitute systems of communication which organizations build up within themselves and which reflect their cultural quiddity: artefacts speak, though we seldom listen, and through them we communicate and act, even if unawares: they have the natural innocence of the physical world, despite the fact that they are inherently artificial…

The production of space

Looking at the floor plans for a new office, we are seduced into believing things will be just as we planned. Every department manager will have inwardly digested the floorplan. Every employee will adhere to the architect's intentions. People will conform to the rules and embrace the corporate vision. But the reality is different. Organisations and employees adapt their work environment to reflect their own identities and their own way of doing things. The 'back office' prevails over the 'front office' as the business of work unfolds.

How do we avoid relying on 'as built' drawings and start to see what is going on in front of us? Just as importantly, how do we go from understanding interactions to understanding relationships – seeing space as more than a floor area on a CAD model? The corporate

Emotion and the symbolic workspace

worth of social capital is greater than physical capital in most organisations. How do we develop a language of space that reflects the importance of social capital?

Lefebvre (1991), a twentieth-century French philosopher, wrote about the 'social production of space' seeking to understand how space evolves (its socio-spatial order). His original work was concerned with the dominance of what he called 'abstract' or 'idealised' space – the type of space represented by maps, plans and models. Designers, planners, managers and architects contributed to the realisation of this 'conceived' space. The utilitarian concerns of 'efficient production' were seen by Lefebvre as being uppermost in their thoughts. He felt that we need to embrace a more complete way of looking at space – one that (a) recognises its evolution over time and (b) acknowledges the role of the individual. Using a spatial triad, three distinct ways of describing space were presented (see also Figure 8.1):

1. *Conceptualised space:* 'Conceived' or conceptualised space reflects the intentions of the designer. It is represented using maps, plans, models and designs – eventually being translated to a physical form. The symbolism present at this level reflects organisational ideologies.
2. *Spatial practice:* Through the physical and day-to-day deciphering of space, individuals develop a daily routine. They adjust to the setting. This is achieved by the "study of natural rhythms, and of the modification of those rhythms and their inscription in space by means of human actions, especially work-related actions" (Lefebvre 1991, p. 117). In a modern context, facilities managers would play some part in modifying

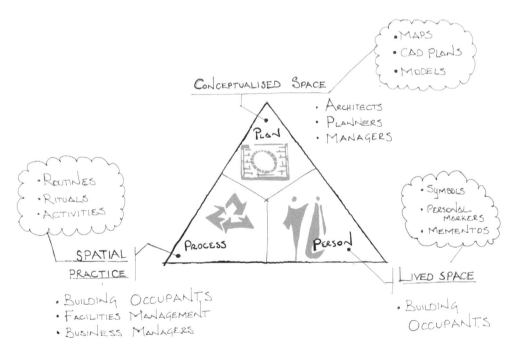

Figure 8.1 Looking at space through a different lens (Lefebvre's triad and the social production of space).
Source: From Lefebvre (1991).

and adjusting routes and networks to support spatial practice. The symbolism embodied at this level reflect team or group ideologies. Spatial practice allows people to 'fit in' and achieve a level of social cohesion.

3 *Lived space*: Represents the perspective of inhabitants and users of space. Lived space overlays physical space, making symbolic use of its objects. It is the passively experienced space which the imagination seeks to change and appropriate (take ownership of). It equates with the personalisation of space in the modern office environment. Lefebvre laments the marginalising of lived space when he suggests that it is "limited to works, images and memories whose content, whether sensory, sensual or sexual, is so far displaced that it barely achieves symbolic force" (Lefebvre 1991, p. 50).

Designers are all too familiar with unintended use patterns. Imposing a 'normative' order on workspace solutions is often futile and undesirable. Space "escapes in part from those who would make use of it" (Lefebvre 1991, p. 26). It seems that the creation of workplace order occurs when people engage with one another and with their setting. How do we make sense of this strange thing we call 'lived space'?

The following sections explore: (1) the use of symbols to represent identity and meaning in lived space; (2) the interpretation of such symbols; and (3) the value placed on symbols by the displayers. We then consider the wider value of symbols as a means of providing social continuity and cohesion.

Workplace identity

The subject of symbols and workplace identity has attracted a lot of interest from researchers in the field of organisational behaviour. Workplace identity can be described as "an individual's central and enduring status and distinctiveness characterisation in the workplace" (Elsbach 2004, p. 100). Workplace identity markers provide a form of expression in much the same way as our choice of clothes, type of car, letterheads and business cards. Despite the widespread use of 'clear desk' policies, many employees continue to personalise their workplace when companies allow it. Approximately 90 per cent of workers in a study undertaken in 2002 were found to personalise their workspace (Wells and Thelen 2002). The same study used factor analysis to identify the most common forms of personalisation (identity markers). Common items included those reflecting personal relationships with family and friends, artwork, musical devices, mementos and trinkets, plants and reading materials.

> I would have plants; I would have a photo or two showing my nieces and nephews when they were younger. My deputy had the same sort of thing. It wasn't mandated by the organisation but was definitely an unspoken agreement that we won't have too much of ourselves, particularly when you were sitting in an open plan space.
> *Interview with consultant using shared start-up space reflecting on experience in corporate work environment*

Liminal space

Betwixt and between

Chloé was feeling left out as a wheelchair user in the office. Despite the best efforts of the space planners to provide an accessible office, the world of business seemed to go on without her. Spaces designated as meeting rooms did not seem to be where decisions were made: they had been made beforehand in the corridors or the washrooms. People appeared to use these 'between and betwixt' spaces to lobby, negotiate or persuade 'off the record'. She could only guess what was being said by her colleagues as she rolled alongside them straining to hear their conversation. It was too late when Chloé entered the meeting room: the decisions had seemingly been made elsewhere.

Transitory spaces in the office are often a fertile ground where emotions, identity and meaning are expressed. As highlighted in the scenario 'Betwixt and between', it is also a place where people feel free to speak 'off the record'. Other 'intimate spaces' go somewhat unnoticed on the official floor plan: corners, secluded spaces and thresholds. Goffman (1959) suggested that people manage their individual identity according to whether they are in a 'public' or 'private' space. In just the same way that actors perform 'front stage' or 'backstage', people in organisations do likewise. They manage how they present themselves according to the audience in front of them. In front-stage regions (front of house), people adhere to certain scripts – they are required to regulate their emotions during interactions with customers, co-workers and superiors. The term 'emotional labour' was used by Hochschild (2012). to describe this 'public face'. In contrast, backstage regions (e.g. restaurants or coffee areas) provide private arenas allowing individuals to step out of character.

Figure 8.2 illustrates a liminal walk-through area with corporate mementos. Liminal spaces represent the in-between spaces that often have a strong symbolic meaning. "Our experiences and how we feel in these spaces are deeply connected to our memories, thoughts and imaginations" (Shortt 2015, p. 636). The term 'liminal space' comes from the Latin word meaning 'threshold' and refers to a period of time/space that is 'in between': it is neither front of house nor back office. It is often seen as an environment that is free from structural obligations and where 'anything may happen' (Turner 2018, p. 13). In this transitory space, people have the opportunity to share secrets and speak frankly. Convention and norms of behaviour are replaced by uncertainty. Not all emotions are necessarily positive. Feelings of anxiety and social separation can emerge when people spend long periods of time in a liminal state (Beech 2011). Corridors are widely used by professionals for daily interactions and communications enabling teaching and knowledge exchange. Its appeal arises because it provides a transitory space without the encumbrance of organisational constraints. However, misuse of liminal space may present ethical issues relating to confidentiality and inclusivity (Hanley 2003).

The presence of liminal space is one of the appeals of the Activity Based Working (ABW) environment with the incorporation of secluded spots, transition spaces and movement between settings. It allows individuals or groups to experience transitions from one state to another. As such, this type of liminal space is something experienced through time as much as through place. Through brief stops, associations with identity, stability

88 Emotion and the symbolic workspace

Figure 8.2 Illustration of liminal space with corporate memorabilia.

and attachment occur. Liminal space becomes the 'lived space' to which Lefebvre (1991) refers. Understanding how to create lived space is the emerging challenge for modern-day office designers.

Investigating a 'crime scene'

We often think of our workspace as a 'crime scene'. A study by Elsbach (2004) observed how office workers used similar approaches to crime scene investigators when evaluating someone else's office. In her study she focused on the interpretation of identity markers by other people (i.e. not the displayer). She wanted to understand the cognitive processes that workers used to formulate opinions about their co-workers based on the inanimate objects found in the workplace. As well as the tidiness or state of neglect, respondents picked up on the type of décor. These included family photos, calendars, formal artefacts, diplomas, ideological artefacts, flashy artefacts and high conformity artefacts. The study provided a fascinating insight into how people formulate opinions about others (even before they have met them) based on the 'crime scene'. Individuals were found to use two strategies:

1 *Top-down profiling:* Likened to the approach used by FBI profilers, this type of evaluation is preferred by time-pressed individuals that rely on a restricted number of salient points to form an opinion. This theory-based approach makes use of past experience to enable individuals to 'stereotype' the displayer whose workplace they are observing. This cognitive method is prone to bias and inaccuracy.

Figure 8.3 Investigating a 'crime scene'.

2 *Bottom-up profiling:* Unlike top-down profiling, bottom-up profiling is data driven and resembles the approach used by police detectives. It relies on a detailed examination of the 'crime scene'. Only the evidence in front of them is considered. It involves greater expenditure of time and a more exhaustive assessment of the evidence. As it does not rely on stereotypes formulated from past experience, it is more likely to be used by a less-experienced assessor. It also leads to less bias in the appraisal process.

This study highlighted how co-workers form initial identity profiles even before an actual personal encounter takes place. Office decoration presents a dilemma for organisational managers seeking to ensure that employees' identities fit with their desired corporate roles. The advent of the non-dedicated office presents yet more dilemmas. When the display of distinctiveness markers is restricted, employees seek out other non-sanctioned behaviours (e.g. putting up permanent artefacts in a non-dedicated office) (Elsbach 2003).

Enclosures

Enclosures or cubicles remain the dominant workspace solution in the USA, contrasting with the widespread use of open plan offices in Europe. By increasing the divider height, the amount of enclosure is also increased. If we were to change the height of the dividers, what impact does this have on: (1) physical outcomes (instrumentality); and (2) symbolic outcomes? Just such a question was pursued by the team at UC Berkeley (Goins *et al.*

90 Emotion and the symbolic workspace

2010). Thirteen sample buildings were used in the study. Two symbolic descriptors – 'homeliness' and 'worker's pride' – were assessed in relation to enclosure height. 'Home' was seen as a common symbol relating to a place of refuge or sanctuary. Physical attributes related to visual privacy and speech privacy were also statistically analysed to see how they were affected by level of enclosure. Privacy reflected a "sense of control over access to oneself or one's group" (Sundstrom *et al.* 1980, p. 102). Figure 8.4 illustrates the use of enclosures in open plan office environments.

The findings of the study suggest that the symbolic attributes (homelike atmosphere and workplace pride) surpassed the physical attributes in terms of perceived impact on worker performance. In fact, they were found to have double the effect of most of the physical attributes. However, increasing the height of the enclosures did not affect the building's symbolic impact. Even the highly valued enclosed office was not seen as a determinant of a homelike atmosphere or feelings of workplace pride. The study concluded:

> ... since the symbolic attributes tested have a larger effect on worker performance, a designer seeking to create a space that maximises worker performance might opt not to include increased divider height in their design in lieu of other elements that support the building's symbolic impact.
>
> (Goins *et al*. 2010, p. 948)

Put another way, the symbolic importance of the workplace is evident, but the answer is not to be found in increasing the amount of enclosure. This is a surprising conclusion given the vehement debate over open plan and enclosed office solutions.

Figure 8.4 Use of partitions in an open plan office.

The IKEA effect

Perhaps you have invested something of yourself in the workplace. Maybe you have chosen your own preferred colour from a template of organisationally approved designs. You might even have been involved in the design or installation process. No matter how small, we like to feel that we have had some input. Just how strong is that sense of attachment arising from our personal investment of emotion? A renowned study demonstrated the symbolic importance of 'self-made' products. The study entitled *The IKEA effect: When labor leads to love* (Norton et al. 2012) involved four studies in which consumers assembled IKEA boxes, folded origami and built sets of Lego. The researchers wanted to examine whether an increase in valuation of self-made products occurred – something they referred to as the 'IKEA effect'. The suggestion is that when people imbue products with their own labour, their effort increases their valuation – giving as an example the 'Build-A-Bear phenomena'.

In their first experiment they attempted to find out whether individuals (DIYers and novices) attached greater value to goods they had assembled compared to similar goods that were preassembled (by an expert). A standard utilitarian (instrumental) product – an IKEA box – was assembled. This did not allow for customisation to reflect the idiosyncrasies of the individual. The experiment was designed to establish the magnitude of the IKEA effect by comparing participants' willingness to pay for their own creations with that of a preassembled item.

A second similar experiment involved a symbolic (hedonic) item that seemed to satisfy the need for self-expression – or showing off to others. The participants were asked to create either an origami frog or crane, and were offered a chance to buy these creations with their own money.

The findings of the experiment confirmed the existence of the 'IKEA effect' for both the utilitarian and symbolic (hedonic) products. People really do believe that their self-made products are as good as the experts'. Interestingly, the study also showed that successful completion was an important part of their increased willingness to pay. People who built and then unbuilt their handiwork (or were not allowed to complete the assembly) did not show an increased willingness to pay. Furthermore, the IKEA effect does not just arise simply because people can incorporate their own idiosyncrasies – it seems that simply 'giving a part of themselves' makes the difference.

This study highlights the importance of coproduction and attachment (McGraw et al. 2003). When it comes to 'workplace chemistry', perhaps we should stop talking about 'users' and start talking about 'coproducers'. Coproduction, participatory design and community-based facilities management have an increasingly prominent role in workplace design and management. It seems that "the evolution in design research from a user-centred approach to co-designing is changing the landscape of design practice as well, creating new domains of collective creativity" (Sanders and Stappers 2008, p. 5).

> Community members do not differentiate between their physical space and their willingness to contribute to it. People do not go, "I like this place therefore I belong to it!" It is not as distinct in people's mind.
> *Interview with consultant in community development*

Our biological Wi-Fi

We are used to creating spaces based on a rational breakdown of work functions. But look carefully and you will see an increasing number of spaces assigned using 'emotional' terminology. Out go words like 'touchdowns', 'meeting rooms' and 'lounges'; in come names like 'snug rooms' and 'home zones' (with soft furnishings to allow confidential meetings) or 'bullpens' to allow people to speak openly. Emotions, it seems, can be matched to the chosen work setting. But designers have little to go on other than intuition. What type of office chemistry works? What type of setting creates a particular emotional response?

So far, we have turned to neuroscience as a possible new lead in our discussion. But just knowing about 'the brain' provides only a glimpse of the mind – its emotions, moods and temperaments. Neurochemistry often provides a more revealing window: the neurotransmitters that flood the entire body. Think of it as our biological Wi-Fi. Neurochemicals quickly alert us to threats or opportunities, and often produce an appropriate response. To understand how neurochemicals affect us in the milieu of the office environment, let's examine one neurochemical: oxytocin. Specifically, how does it affect 'trust' – a socially pivotal emotion in modern-day organisations.

High-trust culture

The level of engagement within organisations is widely acknowledged as being critical to organisational success. When engagement falters the signs are evident in the workplace – people are reluctant to share ideas or information, they neglect mentoring juniors, and there is a general climate of suspicion. The chemistry is not there. Get it right and people start to form strong relationships with colleagues, they feel that they can make a difference, and chances arise to learn and develop. High levels of engagement make people more productive, feel that they have more energy at work, and collaborate better. A 'culture of trust' was found to be the recipe for success in highly engaged organisations (Zak 2017). In a collaborative study on trust (Kosfeld et al. 2005), the team asked the question, "Why do people trust each other?" They undertook an ingenious experiment, having identified a neurochemical called oxytocin as a potential 'trust' chemical. Oxytocin is a neuropeptide produced by the hypothalamus. In rodents it is used to assess whether it is safe to approach another animal that may present a threat. A higher level of oxytocin causes the rodent to approach the other quite willingly.

In the human experiment, a participant was asked to "choose an amount of money to send to a stranger by a computer, knowing that the money will triple in amount and understanding that the recipient may or may not share the spoils" (Zak 2017, p. 87). In so doing the sender was expressing trust. Similarly, if the recipient returned some of the money, they were demonstrating 'trustworthiness'. Before and immediately after this exchange took place, oxytocin levels were taken using blood samples drawn from the recipient's arm. The experiment showed that when more money was received, more oxytocin was produced by the recipient. The level of oxytocin provided a good predictor of how trustworthy a recipient was – or, put another way, how likely the recipient was to share the extra money with the sender.

Several other investigators have examined the social functions of oxytocin. It appears to affect human bonding behaviour, the creation of group memories, social recognition and other social functions (Grinevich and Stoop 2018). But what about the effects of the physical environment on oxytocin levels? After all, if we can create a contagion of the 'love drug' oxytocin in the workplace, we can increase engagement. It seems that it's early days and the most commonly studied environment is the birth environment (Foureur et al. 2010). Oxytocin plays a key part in the female reproductive function, including childbirth and breastfeeding. Another area that has attracted attention in the media and amongst neuroscientists is the practice of having pets in the office. The creation of higher levels of oxytocin produced by interaction with dogs has been associated with higher levels of engagement amongst office workers (Miller et al. 2009, Cunha et al. 2019, Powell et al. 2019). Human–dog interactions increase organisational diversity and the potentially beneficial role of oxytocin in promoting human health.

As an evolving discipline, there remains an ongoing dispute amongst neuroscientists about the relationship between trust and oxytocin (Fujiwara et al. 2012, McCullough et al. 2013). But the link between trust and the positive emotion of 'joy' is not in dispute. Combine the chemistry of (1) trust with (2) a sense of higher purpose, and you are left with a joyful workforce it seems – a correlation of 0.77 (Zak 2017). For companies seeking to increase joy in the office, raising levels of trust may be the answer.

Temperament

What is meant by 'belonging' in the socially diverse workplace? An understanding of temperament can be revealing. In contrast to emotions, temperament reflects your personality – that is to say, you are stuck with it – it is in the genes. In organisations there is a growing awareness that a diverse workforce (including personality) leads to better collaborative working. In an Activity Based Working (ABW) environment, individuals will choose settings that best suit their temperament and prevailing moods. We all have our different ways of doing things. Our work pattern is dictated as much by our state of mind as our calendar entry. Tapping diverse work styles forms the basis of an extensive study by Deloitte, creating a business system called Business Chemistry (Vickberg and Christfort 2017). They teamed up with the biological anthropologist Helen Fisher, whose expertise is on brain chemistry in romantic relationships. They developed a list of traits and preferences by testing and refining three samples involving more than 1,000 professional respondents in each. From this they were able to identify four distinct work styles. Quite unlike conventional personality tests, their four categories were based on neurological studies using fMRI scanners. The purpose was not to 'pigeonhole' employees but to leverage the benefits of a diverse workforce.

Four work styles (personality traits) were identified in the study (Brown et al. 2013). These were associated with four neurochemical systems:

1 *Dopamine system:* These subjects tended to be creative, energetic, possess a higher level of curiosity and possess mental flexibility. They pursue risk and novelty.
2 *Serotonin system:* Amongst these subjects there is a greater drive to belong and be sociable. Their approach tends to be fairly traditional with a reluctance to explore.
3 *Testosterone system:* These subjects have a preference for 'rule-based' thinking (engineers, computer scientists, maths and music). They tend to be decisive, sceptical and assertive.
4 *Oestrogen/oxytocin system:* People in this category show a preference for intuitive and imaginative thinking. They tend to be long-term thinkers. Having good verbal and social skills, they are particularly trusting and empathetic.

Imagine being a space planner equipped with this knowledge about work-style neurochemistry. A person for whom the serotonin system is dominant (risk averse) might have difficulties working alongside a person for whom the dopamine system prevails (risk taker). For those that like spontaneity, the behaviour of the detailed thinker would be exasperating. Yet through careful planning, managers can benefit from these team differences. When it comes to designing space, it points to a diverse solution to satisfy the demands of a truly diverse population (including temperament). The designer suddenly becomes an alchemist!

Summary

This chapter has illustrated how meaning and attachment affect our daily workplace experience. Symbolism provides the opportunity for organisations and individuals to communicate emotions. However, people may not interpret the symbol in the way that it had been intended by the 'expresser'. In our quest for a 'non-stick' plug-and-play territory, employees are forced to use ever more ingenious ways to convey their individuality. Developments in Activity Based Working offer the opportunity to embrace an emotionally diverse workforce. However, tailoring the various settings to satisfy the varied temperaments is more of an art than a science. Understanding office space as 'lived space' challenges the dominant approach to space planning. We need to harness scientific know how in order to avoid a 'blunderbuss' approach to workplace engineering.

In the final conclusive chapter, we bring together the three distinct threads of instrumentality, aesthetics and symbology in the workplace. Pointing to future directions, it suggests that a flourishing and resilient workforce can be advanced through good workplace chemistry.

References

Beech, N., 2011. Liminality and the practices of identity reconstruction. *Human Relations*, 64 (2), 285–302.

Brown, L.L., Acevedo, B. and Fisher, H.E., 2013. Neural correlates of four broad temperament dimensions: Testing predictions for a novel construct of personality. *PLOS ONE*, 8 (11), e78734.

Bruce-Mitford, M., 2008. *Signs & Symbols: An Illustrated Guide to Their Origins and Meanings*. London: Dorling Kindersley.

Campbell, J., 2002. *The Flight of the Wild Gander: Explorations in the Mythological Dimension Selected Essays 1944–1968*. Novato, CA: New World Library.

Cunha, M.P., Rego, A., and Munro, I., 2019. Dogs in organizations. *Human Relations*, 72 (4), 778–800, 0018726718780210.

Elsbach, K.D., 2003. Relating physical environment to self-categorizations: Identity threat and affirmation in a non-territorial office space. *Administrative Science Quarterly*, 48 (4), 622–654.

Elsbach, K.D., 2004. Interpreting workplace identities: The role of office décor. *Journal of Organizational Behavior*, 25 (1), 99–128.

Foureur, M., Davis, D., Fenwick, J., Leap, N., Iedema, R., Forbes, I. and Homer, C.S.E., 2010. The relationship between birth unit design and safe, satisfying birth: Developing a hypothetical model. *Midwifery*, 26 (5), 520–525.

Fujiwara, T., Kubzansky, L.D., Matsumoto, K. and Kawachi, I., 2012. The association between oxytocin and social capital. *PLOS ONE*, 7 (12), e52018.

Gagliardi, P., 2011. *Symbols and Artifacts: Views of the Corporate Landscape*. Berlin: Walter de Gruyter.

Goffman, E., 1959. *The Presentation of Self in Everyday Life*. Oxford: Doubleday.

Goins, J., Jellema, J. and Zhang, H., 2010. Architectural enclosure's effect on office worker performance: A comparison of the physical and symbolic attributes of workspace dividers. *Building and Environment*, 45 (4), 944–948.

Grinevich, V. and Stoop, R., 2018. Interplay between oxytocin and sensory systems in the orchestration of socio-emotional behaviors. *Neuron*, 99 (5), 887–904.

Hanley, J., 2003. Ignorance of the law excuses no man. *Directions in Psychiatry*, 23 (Lesson 12), 151–158.

Hochschild, A.R., 2012. *The Managed Heart: Commercialization of Human Feeling*. Oakland, CA: University of California Press.

James, W., 2013. *The Principles of Psychology*. Redditch, UK: Read Books.

Kosfeld, M., Heinrichs, M., Zak, P.J., Fischbacher, U. and Fehr, E., 2005. Oxytocin increases trust in humans. *Nature*, 435 (7042), 673–676.

Lefebvre, H., 1991. *The Production of Space*. Oxford: Blackwell.

McCullough, M.E., Churchland, P.S. and Mendez, A.J., 2013. Problems with measuring peripheral oxytocin: Can the data on oxytocin and human behavior be trusted? *Neuroscience & Biobehavioral Reviews*, 37 (8), 1485–1492.

McGraw, A.P., Tetlock, P.E. and Kristel, O.V., 2003. The limits of fungibility: Relational schemata and the value of things. *Journal of Consumer Research*, 30 (2), 219–229.

Miller, S.C., Kennedy, C.C., DeVoe, D.C., Hickey, M., Nelson, T. and Kogan, L., 2009. An examination of changes in oxytocin levels in men and women before and after interaction with a bonded dog. *Anthrozoös*, 22 (1), 31–42.

Norton, M.I., Mochon, D. and Ariely, D., 2012. The IKEA effect: When labor leads to love. *Journal of Consumer Psychology*, 22 (3), 453–460.

Powell, L., Guastella, A.J., McGreevy, P., Bauman, A., Edwards, K.M. and Stamatakis, E., 2019. The physiological function of oxytocin in humans and its acute response to human–dog interactions: A review of the literature. *Journal of Veterinary Behavior*, 30, 25–32.

Preiser, W.F.E., 1995. Post-occupancy evaluation: How to make buildings work better. *Facilities*, 13 (11), 19–28.

Sanders, E.B.-N. and Stappers, P.J., 2008. Co-creation and the new landscapes of design. *CoDesign*, 4 (1), 5–18.

Shortt, H., 2015. Liminality, space and the importance of "transitory dwelling places" at work. *Human Relations*, 68 (4), 633–658.

Spenser, E., 2007. *The Faerie Qveene*. London: Pearson Education.

Sundstrom, E., Burt, R.E. and Kamp, D., 1980. Privacy at work: Architectural correlates of job satisfaction and job performance. *Academy of Management Journal*, 23 (1), 101–117.

Turner, V., 2018. *Dramas, Fields and Metaphors: Symbolic Action in Human Society*. Ithaca, NY: Cornell University Press.

Vickberg, S.J. and Christfort, K., 2017. Pioneers, drivers, integrators and guardians. *Harvard Business Review*, 95 (2), 50–56.

Wells, M. and Thelen, L., 2002. What does your workspace say about you? The influence of personality, status and workspace on personalization. *Environment and Behavior*, 34 (3), 300–321.

Zak, P.J., 2017. The neuroscience of trust. *Harvard Business Review*, 95 (1), 84–90.

Zeisel, J., 1984. *Inquiry by Design: Tools for Environment-Behaviour Research*. Cambridge, UK: Cambridge University Press.

9 Conclusions

The offices of today are almost unrecognisable from the incarnations of yesteryear. The connection between organisations and employees has shifted; the expectations of employees have changed; the social contract has been redefined. Relationships with organisations have become short lived, as have relationships with other employees. The typology of the workplace has changed so that the starting point is the city rather than the office, as pointed out by Laing (2014, p. 11):

> Two big shifts stand out which have more general implications for work and place: a shift towards collaborative and urban 'workscapes' that are more heterogeneous, mixed-use and multi-scaled; and a related shift to the collaborative consumption of workspace and workspace-as-service.

As well as co-working arrangements, other communal models are emerging such as 'open house', 'third places', 'pop-up places' and 'interstitial spaces'.

Many of these typologies are characterised by impermanence. Yet at the same time they provide opportunities for collaboration that conventional 'bricks and mortar' cannot. In fact, they appear to offer the capacity to create social capital in a more fluid, responsive landscape.

Against this backdrop, we consider the human condition. It has the capacity to be overwhelmed by change and 'future shock'. But at the same time, it can be liberated and empowered by the untethered workplace. Negative emotions may arise in a stressful fast-changing technological world, yet positive emotions may also find breathing space.

We might be inclined to dismiss people's emotional resistance to change. It might be perceived as a generational failing – you are just getting too old! However, during the course of writing this book we have had extensive conversations with designers, managers and occupiers in order to make sense of this world. We found that emotions formed a key role in everyone's relationship with the workspace. We also found that it acted as an important mediator influencing human interactions. People were more inclined to express emotional responses to office space as a tool or enabler. They were more reluctant to express aesthetic or symbolic sentiments. This was either because such expression was socially less acceptable or because they experienced a 'framing' problem (i.e. they were not able to identify or attribute emotional experiences to their physical surroundings).

In defining the 'emotionally intelligent building', we explored its counterpart – the 'intelligent building'. It was found that the technological focus of intelligent buildings reflected a preoccupation with the utilitarian or performance-based aspects of the building.

Emotion tended to be marginalised with a focus on quashing negative emotions – or making miserable people less miserable.

Our findings consistently showed that an instrumental outlook on workspace design was 'necessary but not sufficient' to create an enduring work environment. Two prerequisites were the aesthetic and symbolic dimensions. These dimensions were required in order to create flourishing and social individuals. Good design is much more than ornamentation: it reaches every aspect of the human 'psyche'. But unlike Sullivan, whose maxim was 'form follows function' (1896), the expression 'form follows emotion' seemed more apt. Only by addressing the aesthetic and symbolic characteristics of the workplace is it possible to realise its instrumental possibilities – taken together, these two characteristics provide the motive or driving force. A preoccupation with function in modern workplace design marginalises 'lived space': it leads to a technological bias.

Space planning has evolved in tandem with office innovations. The adoption of social network analysis and accompanying tracking technologies has created new insights. The 'mass data' created from our 'digital exhaust' tells us when space works and when it does not. It throws light on how people choose to use the space and how we engage with it.

Emotion remains a problematic concept – it gives rise to some divergence and convergence of opinion between scientific disciplines. Neuroscientists, linguists and social scientists often fail to agree on terminology and mechanisms. This is evident in the various research findings detailed in the preceding discussions. It is an elusive field and yet we acknowledge its pivotal importance in modern business. This book's aspiration was to uncover the dimensions of the 'emotionally intelligent workspace'. It does not attempt to provide a prescriptive, rating-based system for work environments. It (unapologetically) raises as many questions as it answers – but it does attempt to show the cavernous gap in our understanding of emotions and their spatial context. The discussion has covered the following stages (using the REACT model (MacCrimmon *et al.* 1986)):

- recognise;
- evaluate;
- adjust;
- choose;
- track.

Recognise: What are the affects (temperament, mood and emotion) that play out in the office environment? What are the mechanisms underlying each (hygiene factor, sensory or memory)? Understanding the 'framing' problem is the first challenge in creating the emotionally intelligent workplace.

Evaluate: Perhaps one of the greatest challenges – how do we measure it? Language presents us with many varied and nuanced ways of expressing emotions. Linguistics and psychology give us robust ways to interpret people's expressions (including facial, physiological and behavioural). Developments in neuroscience including fMRI and EEG provide us with additional insights (although often overpromising on their ability to discriminate between emotions).

Adjust: Can we influence emotions by adjusting the workplace setting? What are the office parameters that influence workers' ability to flourish? Many environmental factors including acoustics, lighting, nature and colour were briefly considered. However, it was noted that adjusting one factor in laboratory conditions often proved misleading. Only by considering the combined effect of environmental factors can the designer get close to understanding their influence on emotions.

Choose: We are bombarded with choice in the modern workplace. This is particularly evident in the Activity Based Working (ABW) environment. Choices about where, when and how we work can be both liberating and challenging at the same time. Given freedom of choice, office occupants often make the wrong choices. How can we design an environment that encourages people to choose the right work setting? Issues such as switching behaviour play a pivotal role in making Activity Based Working (ABW) work in practice. The application of 'nudge' and the designer's role as 'choice architect' may provide some answers, taking on responsibility for organising the context in which people make decisions.

Track: Achieving the right office chemistry is not a 'one-shot' process. The office is a dynamic 'lived space' rather than a 'conceptualised space'. It fulfils the needs of a diverse workforce, possessing a multitude of temperaments and work styles. Technological innovations using tracking technologies and social network analysis (SNA) may provide some day-to-day analytical tools. However, going from understanding 'interactions' to understanding 'relationships' remains the all-important challenge. Incorporating measures of 'emotional intelligence' as part of a post-occupancy appraisal provides a resourced mechanism for monitoring the success of a design solution.

Reflections

> You can discover more about a person in an hour of play than in a year of conversation.
>
> (Lingard 1670)

This quote can aptly be used to explain the lasting appeal of the office despite its doomsayers. Face-to-face interaction allows us to exchange ideas, loyalties and trust in a way that virtual communication cannot. But in order to leverage the full potential of the office, many designers and managers are looking beyond the instrumental horizon. Aesthetics and symbolism allow us to make sense of the past and present: they are not ornamental luxuries but a necessary characteristic of the contemporary workspace. As noted by the novelist Renee Dubos, our workspace should provide:

> …reverence for the past, love for the present, and hope for the future.
>
> (Dubos, 1965, p. 279)

References

Dubos, R., 1965. *Man Adapting*. New Haven, CT: Yale University Press.
Laing, A., 2014. Work & place. *Occupiers Journal*, 3 (1), 11–14.
Lingard, R., 1670. Letter of advice to a young gentleman leaving the university concerning his behaviour and conversation in the World.
MacCrimmon, K.R., Wehrung, D.A. and Stanbury, W.T., 1986. *Taking Risks: The Management of Uncertainty*. New York: Free Press.
Sullivan, L.H., 1896. *The Tall Office Building Artistically Considered*. Los Angeles, CA: Getty Research Institute.

Index

Page numbers in **bold** denote tables, those in *italics* denote figures.

5S system 60–61

ABC of psychology 1, 6, 9
Activity Based Working (ABW) 65–66, 68, 94, 98; environment 66, 87, 93, 98
activity trackers 39
aesthetic: appeal 75, 78; assets, emotional triggers arising from 52; beauty, measuring 78; designer focus on 50; features of a workspace 72; machine 22–23; mimicry 76; qualities of nature 75; responses evolve in 'lived-in' space 79; value of naturalness 77; ways of working 61; workplace 72, 74, 97; worldview 58
aesthetic dimension: influence response to a scene 78; prerequisite for enduring work environment 97; self-actualising properties of 76; in sensemaking 51–52, *53*, **54**, 56, 74
aesthetic emotions 74–75; judgements arising from 74; reluctant to express 96; triggers arising from 52
aesthetic experience 54; of built environments 79; innate and spontaneous 84; seen as marginal to practice of business 80
aesthetics 51, 74, 83, 94, 98; environmental 72; mechanisms influencing 53; neuroaesthetics 77–77; related to satisfaction 52; support individual and organisation 56; used in sensemaking 55
affect 1, 6, 9, 14, 18, 21, 68; ability to generate positive 47; of emerging issues on work styles 36; of environmental factors on experience of space 80; evolutionary path of building according to 38; of level of enclosure 90; of meaning and attachment on daily workplace experience 94; measures of 47; of neurochemicals in office environment 92; in office environment 97; of stimulating environment on mood 46; of workplace on everyday well-being 72; of workspace on new type of employee 25
affection 26, 28

affective features of the workspace **48**, 58; interaction 46; qualities of place/space *18*, *19*; responses to place 18; state, impact of different interior forms on 79; states 17
alexithymia 44
Allen, T.J. 39, 61, 63
ambient intelligence 39–40
anger 10, 14, **54**, 58; about encroachment in the workplace 67; arises in situations of threat or thwarted goals 67; central role in influencing our experience of infringement 67; determinants in territorial workplace 67; frustration leads to 46, 70; over infringement of territory 68; linked to territory 67; sub-threshold version of 17
Appleton, J. 75
Arthur, C. 43
Attention Restoration Theory (ART) 77

Bakker, I. 19, 74
Bar-On, R. 47
behaviour
Biophilia Hypothesis (BH) 77
Brown, G. 59, 67
building automation and control systems (BACS) 37
building automation systems (BAS) 35, 37
building intelligence 3; can undermine emotional intelligence 2; ideas starting to emerge 24; indispensable part of modern work environment 34; meaning more technology 41; overrated 45
buildings 34–36, 40; accommodate new technology 36; blinkered behavioural approaches to design 48; design, hierarchy of needs related to 26; emotional 4; emotions marginalised 96–97; evolutionary paths according to affect, behaviour and cognition 38; flexible 34, 36, 40; green 34, 37; high-tech 34, 37, 40; impact of 'naturalness' 77; interoperable 34, 37, 40; IQ, facilities

buildings *continued*
  managers encouraged to know 36; new technology in 41; qualities attached to physical surroundings 17; symbolic impact of 90; tall, perspective in 22; technology 2–3, 35; technology-laden 3, 32, 36; tracking evolution of intelligence 31, 34; unable to accommodate demands for new technology 36; users, emotional responses of 4; workplace automation systems 35; *see also* emotionally intelligent building, intelligent building (s)

California Management Review 52
challenge: blinkered behavioural approaches to building design 48; of diversity 8; key 6; low occupancy levels of dedicated spaces 60; of man-made environments 79; measurement 79, 97; modern office 3; of new working practices 8; seeking out new 25
challenge to designers 17, 60, 88; in creating emotionally intelligent workplace 97; to design an environment of coproduction 7; to make space 'readable' 65; to understand office space as 'lived space' 94; understanding 'relationships' 98
Clements-Croome, D. 35, 38, 41
cognition 1, 6, 9, 38, 54, 73
communicate/communicating 54; emotions through symbolism 94; less with distance 39; through artefacts 84; wireless systems 39
communication 1, 65; business, devices to support 34; complex, allowed by symbols 84; face-to-face and digital, diminishing with distance 39; face-to-face better than virtual 98; integrated systems (ICS) 37; intensely networked 73; office-wide, improve 70; physical space as vehicle for 84; standards, absence of 35; tool, office as 43; wireless Wi-Fi 6, 37; workplace 30
communications: corridors widely used for 87; electronic 35; employees heavily dependent on 66; hardwired 36; mapping 40; *see also* telecommunications
coproduction 7, 91
core intelligences: perceptual/organisational and verbal/propositional 45
co-workers 7; co-located 39; form initial identity profiles before personal encounter 88–89; interaction with 17; need for relationships with 30; regulate emotions during interactions with 87
co-working 7, **48**; arrangements 96
'crime scene' 88–89; investigating 89
culture 21, 83; consistent thread of office in 24; customised designs to meet unique requirements of **48**; designers interested in 49; dual 63; emotions arising from symbolic

dimension reflect variations in **54**; high-trust 92; organisational 7–8; trust-based 1

demands: of clients, changing 5; of diverse population 94; environmental 47; of hot-desking 59; of owner, initial and changing 38; meeting 5
demands of modern-day offices 4; cooling 37; for new technology, buildings unable to accommodate 36
distrust/mistrust 59, 79
Dubos, R. 55, 98
Duffy, F. 4–5, 33, 45

EEG (electroencephalography) 78–79, 97
electrodermal response sensor (EDRS) 40
Elsbach, K.D. 52, 59, 86, 88–89
emotional 25; abilities 47; anatomy/infrastructure 12; awareness 44; competence 46; design palette 54, 55; disabilities 45; dysfunction 47; element present in consumer products 23; engagement of employees failed 9; facet of negative emotions 56; feedback 55; inadequacies of employees 47; information 2, 45; interpretation of symbols 84; judgements, aesthetic 74; knowledge 46; labour of public face 87; mechanisms/triggers 52; needs 3; payoffs, dual 23; potential 4; regulation 40; resistance to change 96; setting 4; shortcomings 44; understanding of users, technology falls short of 40
emotional building 4; impact of 'naturalness' 77; qualities attached to physical surroundings 17; tall, perspective in 22
emotional factors in workplace: denial of 15; environments, quotient of 47; qualities of 42; well-being influenced by 47
emotional language/expression/terms 46; marginalisation of empathy 48; not tolerated at work 72; positive 51; signals 46
emotional states: affect worldview 14; current (affect) 21; at work 74
emotional intelligence (EQ) 2–3, 6, 41, 43, **44**, 45, 47, **48**; ability based/mixed based 47; of a building 42; dominant models for evaluating 45; elusive third intelligence 45; evidence-based approach to 3, 42; glue for organisations 44; harnessed 56; instrumentality, aesthetics and symbology dimensions to characterise 51; interpretations of 47; interventions to enhance 4; as measurable phenomenon 45; measuring with ability model 46; overlaps with cognitive ability 46; as part of post-occupancy appraisal of office space 98; predictor of performance/success at work 3, 45; seemingly intangible characteristics of 45; set of interrelated abilities to exercise 46; trait

EI 47; of work environment 47; of workplaces 45
emotionally intelligent building 2–3, 32, 44, 47; capacity to accommodate changing human needs 6; defining 42, 96; to investigate 34; provides 'intelligence' in feedback or cues 49; remote working misses out on something offered by 43; thwart 4
emotional responses 51; of building users 4; to colour of fleet of buses 51; dimensions to capture 52; evolve in 'lived-in' space 79; involves a number of underlying stages 10; to making sense of infringement 67; to office space as tool or enabler 96; of people to surroundings 18; range of, to three dimensions of any artefact 54; research in highly controlled conditions 74; sensemaking process produced 52; sensory modes combine to produce 80; setting creates a particular 92; in social settings, unempathetic and ineffective 44; stages 10, *11*; unforeseen 52; by users in the workplace 61
emotion (s) 1, 15, *16*, 93, 97; ability to reason about and use 46; aesthetic 74–75; affect how and what we see 10; anger linked to territory 67; appeal of form/design to 22–23; associated with aesthetic dimension 54; and brain *12*, 92; chief function of 10; communicated by symbolism 94; detection 40; different 18; different, discriminate between 2; different, experienced during the day 17; duration fleeting/short-lived 6, 17; elusive nature of 60; emotional intelligence ability with 4; enable animals to function as part of social system 12; evoked by symbols 84; experts/researchers 46; form follows 22, *23*; free 23; identifying and describing 44; imperative voice of 22; -inducing quality of a space 18; influencing 4; innovations relying on 68; instrumental 75; managed/mismanaged in pursuit of company goals 46; marginalised in emotionally intelligent buildings 96–97; mechanisms that evoke 52; monitor 2; mood as sub-threshold counterpart to 17; multistage process of experience 19; nature of 11, 19; neuroscience discriminates between 97; and nudge 63; towards the organisation 52; of others, appreciating 44; perceiving 46; pivotal contribution in organisations 6; power of 2; precarious relationship with technology 31; reductionist explanation of 79; reflect current emotional state/affect 21; reflect variations in culture and past experiences 54; SADNESS 13; science of 8–9; SEEKING 13; signals conveyed by 46; spread in a two-dimensional space 18; symbolic 75; triggered by objects rather than outcomes 75; trust, socially pivotal in modern-day organisations 92; types of 11, 58; unpredictability of 15; utilitarian 74
emotions arising: from ambiguity of ownership in shared space 68; through disruptive process mediated by judgement/cognition 73; from functional/instrumental dimension 53; from symbolic dimension 54
emotions, control/regulate/regulation 12; ability to 47; containing 72; required during interactions with others 87
emotions and environment 17, 21, 80; enables function and survival in response 10; influence of environmental factors on 97–98; office environment can stimulate 17; physical environment used to manage 4, 6; in shared environment 61
emotions, expressing 4, 47; or containing 72; in transitory spaces in offices 87; using physical environment 4; ways of 97
emotions, human 2; connection with nature 77; opportunity to engage with 3; representation of 19
emotions, negative 14–15, 19, 51, 53, **54**, 56, 58, 74, 96–97; quashing 58, 97; in stressful fast-changing technological world 96
emotions and motivation 10, 25; interaction 8
emotions, personal: investment of 91; recognition and control of 47
emotions, positive 13–15, 19, **54**, 74, 93, 96; aesthetic dimension invites more 74; evoked by interaction with physical environment 80; experiencing 14, 25
emotions, roles/part played: in achieving instrumental goals in the workplace 63; in instrumental, aesthetic and symbolic dimensions 52; in instrumental worldview 58; in sensemaking 50–51; in world of technology 58
emotions, understanding 9, 11, 17, 21, 25, 46–47, 66, 78; gap in 97; symbolic 68
emotions in workplaces 9, 21; associated with getting work done 70; attempts to suppress 15; key role in relationships with 96; make us function 23; matched to chosen setting 92; prevalence of 59; role played in achieving instrumental goals 63
employees: banned from working remotely 43; changed expectations of 96; conflict and ambiguous boundaries between 61; decide where/how to work 65; digital exhaust 40; with diverse activity profiles/work styles 66, 93; emotional inadequacies of 47; enticing new 3; feel ownership over territories 67; forced to use ingenuity to express individuality 94; insecurity of 46; maverick, feel dispossessed 63; need magnet to draw them to work 43–44; negotiate nuanced

Index

employees *continued*
  world around 64; in non-dedicated office 89; organisations haemorrhaging 9, 44; paid for output 55; reduced travel time for 7; relatedness needs 30; satisfaction from ABW environment 66; settlers or vagrants 63; studies on 65; study of sensemaking in 51; used SSD at standing height 69
employees' relationships with one another 40; shorter 96; undermined by hot-desking 59
employees' workplace: capacity to re-engage with 9; enables getting the job done 54; personalise/adapt environment 84, 86; response determined by perception of safety 27; visibility across office space 46
ERG theory (existence, relatedness, growth) 30
evaluating emotional intelligence: ability model 45–46; mixed model (traits and abilities) 45–47
evolution: of the brain 12; in design research 91; electronic 24; of the intelligent building (IB) 31, 34, 38; of the office 24; of space 85

facilities managers 1, 3–4, 15, 35, 43, 50; alert for signs of dissatisfaction 58; allocate space 39; encouraged to know building's IQ 36; post-occupancy evaluation 84; role in shared environments 61; role in spatial practices 85–86
flexible: office 59; partitions and furniture systems 33; working 61, 72; workspace 68
flexible building 34, 36, 40
form follows function 21–22, 97
Fredrickson, B.L. 14
functional MRI (fMRI) 78–79, 97; scanners 93

Goffman, E. 73, 87
Goins, J. 89–90
green building 34, 37
green bus 51–52, 60

habitat theory (Appleton's) 75
Harvard Business Review 72
heart rate sensor 40
heating 22, 35, 70; control panels 70
Herzberg, F.I. 28, 58
hierarchy of needs 31; Maslow's 25–26, 30; related to building design 26
high-tech building 34, 37, 40
Hirst, A. 63, 73
Hoendervanger, J.G. 66
hot-desk/hot-desking 28–29, 59; nomads 63; open plan work environment 4; policy 46, 61; space savings made possible by 65
human capital 6–7
human potential 4, 25, 40

HVAC (heating, ventilation and air conditioning) 35, 37
hygiene factors 28, 53, 58, 97; perspective 70; in shared environment 60; workspace instrumental dimension **54**

insecurity of employees 46, 72
instrumental 98; approach to fitting a person to a workstyle 61; assets, emotional triggers arising from 52; emotions 75; goals in the workplace, achieving 63; interpretations of contemporary multiple-activity work settings 49; objects/space 68; outlook on workspace design 97; perspective 48, 58, 66, 70, 73; possibilities realised 97; possibilities of space 48; product, utilitarian 91; view of office design 40; view of workspace **48**; workplace 58, 68, 72
instrumental dimension: of an artefact 53; negative emotions of 56, 68; of sensemaking of artefacts 51–52; tied to levels of dissatisfaction 53; of workspace characteristics **54**
instrumentality 48, 51, 55, 60, 94; indicator of employee satisfaction and effectiveness 52; mechanisms influencing 53; height of office dividers, impact on 89; of workspace 54
integrated communication systems (ICS) 37
intelligent building (s) 2, 44; able to harness intelligence from every device 37; attempts to create 4; attempts to leverage power of technology 2; centralised 39; conceived in first wave of knowledge economy 6; conceived as technical fixes 37; concept of 42; concerns of privacy and control 34; defined by ability to accommodate new technology 36; definition expanded to encompass users 37; design characteristics as important as technology in realising 36; emergence of 34; emergence thwarted 35; early formulations of 3; enabled office users to engage with computers/other workers 3; evolution of 38; flexible enough to accommodate changing needs 36; flexible and interoperable 40; growing discontent with technological definition of 37; obsolete and unintelligent buildings could now become 37; response to ad hoc arrival of new technologies 32; successes and limitations of 32; success of remote working 43; technological focus of 96; technology-laden 3, 32; *see also* emotionally intelligent building
interoperable building 34, 37; intelligent 40

James, W. 84
job demands-resources (JD-R) model 59

Karandinou, A. 78–79
Kellert, S.R. 75, 77
Kenrick, D.T. 26, 30
know how 6, 15; scientific 94
knowledge economy 6

lean thinking 60
Lefebvre, H. 85, 86, 88
Leonardi, P. 40
leverage attempts to 2; benefits of a diverse workforce 93; full potential of the office 98; human input 31
leveraging human potential 4; power of emotion 2
Lowe, G.S. 24

McGurk, H. 74; McGurk effect 74
magnetic resonance imaging (MRI) 78
Maslow, A.H. 25–26, *27*, 28–30
Maslow's hierarchy of human needs 25–26, *27*, 28, 30
maverick working 63
Mayer, J.D. 46–47
Mayer–Salovey–Caruso Emotional Intelligence Test (MSCEIT) 46
Montag, C. 11, 13
mood 6, 9–10, 15, *16*, 17, 19, 21, 92; affected by environment 46; choose settings that best suit 93; improved by organic workplaces 76; in office environment 97
Morrison, R.L. 59
motivate 10, 21; self-motivated 47
motivational: predisposition 10; systems have dedicated need-state detectors 13
motivation change in 11; emotions interact with 8; intrinsic 25, 30; stimulating environment can dramatically affect 46; tools provide basic insights into 40; understanding 25

neuroaesthetics 77–79
neuroarchitecture 77–79
neurological differences between three types of affiliation 28; identification of four work styles 93; research 9; work on deep brain stimulation 11
neuropsychology, emerging ideas in 15
neuroscience 77, 92; developments in 78, 97; popularisation of 79; providing new insights about environmentbehaviour interaction 79
neuroscientific evidence 79
New Era of Positive Psychology, The 14
Newport, C. 1, 31
new technology 58; ability of intelligent buildings to accommodate 36; appearing in workplace 35; in buildings 41; change in thinking on design 33; demanded by business for workplace 36; enabled convergence, miniaturisation, mobilisation and modularisation 32; hampered productivity 32; instrumental workplace transformed by 68; more efficient through 25; office scenery designed to assimilate 35; rate of change imposed by 40; stands in way of person-to-person/person-to-environment interaction 32; using appeal of 3
nomad/nomadic workers 62–63; *see also* maverick working
non-stick environment 4, 94
non-territorial office 59, 66
nudge 56, 65, 69–70, 98; default nudge 69–70; relies on emotion and behaviour change 63; technique used to change behaviour at work 68

office environment (s) 1, 12, 49; affects of temperament, mood and emotion play out in 97; basic human needs taken for granted 27; can be used to stimulate emotions 17; envisaged, radically different 65; introducing 5S system 60; mood more commonly experienced 17; neurochemicals affect 92; nudge used to address 70; open plan, use of enclosures in 90; range of feelings played out in 16; role in motivation at work 21; seen as tool for achieving key behavioural/economic goals 48; shared, levels of distraction reported 59; to stimulate positive emotion 14
office environment (s), modern 13, 60–61, 86; devices used to encourage SEEKING behaviour 13; office etiquettes introduced 61; personalisation of space in 86; tension exists in 60
office etiquette 60–61, 73
office innovations 40, 97; Activity Based Working (ABW) 65, 68; ad hoc 35; individual 35; recent 56; technological 98
office landscape 8, 24, 54
office workers 1, 25, 30, 35, 59–62, 66; having pets in the office 93; Japanese, death by overwork 72; workspace as 'crime scene' 88
open plan 73; areas 63; design office 64; designs 68; environment 13; environment, technology-laden 65; hot-desking work environment 4; landscape 35; office environments 90; offices 29, 63, 73, 83, 89, 90; sharing environments 60; space 13, 16, 83, 86; work environment 28
Orbit report 36; ORBIT-2, 36
organisational: approval of designs 91; artefacts, sensemaking of 51; behaviour, researchers in 86; benefits of teleworking 59; brain, hook up with 25; change 4, 36; concerns about security, productivity and waste 62;

organisational *continued*
  constraints 87; culture 7–8; diversity increased 93; expression through symbolism 56; goals 75; ideologies reflected in symbolism 85; intelligence 45; literature, sensemaking in 50; office decoration dilemma for 89; research 50; style 61; success, engagement critical to 92
organisational performance 39; stimulated by nudge 56
organisational workspace: drivers of reinvention of 9; instrumental 58; outcomes improved by affective aspects 58
organisations 24; accumulate social capital 6; adapt work environment to reflect identity 84; aesthetics support 56; attempting to create 'trust-based' culture 1; benefits of hot-desking 59; contribution of people in the workplace 9; corporate worth of social capital 85; creative 54; dead spaces not being used 66; diverse workforce leads to better collaborative working 93; emerging 7; employee relationships with 96; fit of emotionally intelligent work setting with 55; focus on change and output 54; haemorrhaging talented employees 9, 44; hot-desking 28; importance of relatedness needs 30; increasingly aware of the 'pull' of the work environment 14; instrumental objects including space 68; introduction of open plan offices 63; investing in the sitstand desk (SSD) 69; level of engagement within 92; low-change/routine contrasted with high-change/non-routine 36; making sense of 49; need to address technology problems 35; people interactions 40; people manage individual identity 87; personalisation of shared workspace 86; pivotal contribution of emotions in 6; presenteeism 72; promoted as natural and environmental 51; resort to using headphones 4; sensemaking process evoked emotion towards 52; socially pivotal emotion of trust 92; space needs 36; successful, importance of relationship chemistry to 3; support Activity Based Working 65; trend for coworking 7; use to convey ideologies and structures 84; vital part of face-to-face interaction in 1; withdrawal of commitment to 73
organisations, communication in 43; emotions in symbolism 94; intense networking needed 73

post-occupancy evaluation (POE) 9, 39, 63, 84
Pratt, M.G. 17–18, 49
Preiser, W.F.E. 63, 84
prompt/prompted by 10, 14, 60, 65, 75; of anger 67–68; emergence of 'smart energy grids' 37; studies 51–52; for workplace designers 68; *see also* nudge
psychological walls 63–64
psychology ABC of 1, 6, 9; emerging ideas in 15; evolutionary 13; gives robust ways to interpret people's expressions 97; healthy half of 25; neuroscience supporting evidence from 79; positive 13–14, 44, 70

Rafaeli, A. 51–52, 53, **54**
recognition 28; of blurred boundaries **48**; of emotion 47; face 40; social 92; systems change beyond 24
resources 9; access to 58; arising from hot-desking strategy 59; focused on satisfying greatest need 26; good fit between employee, job requirements and 61; human 62; internal, drawing on 11; limited cognitive 77; manager, human 43; mental, drained 27; personal 14–15; physical and cognitive, drain on 28; restore attentional 77; SEEKING to secure 13; sharing 28; social 15; type needed 62; working time spent looking for 60
Russell, J.A. 17, *18*, 19

Sanders, E.B.-N. 65, 91
scenery 33, 35
security ensured by ambient intelligence 39; basic human need for 26; mechanical and electrical systems required to maintain 33; organisational concerns about 62; threatened feelings of 28
Seligman, M. 13
Sennett, R. 29, 72
sensemaking 65; enabled by symbolic aspect of workspace 83; in office landscape 54; of organisational artefacts 51; process produced an emotional response 52; properties of an environment 49; result of interaction with space and artefacts within office space 55; roles played by emotion in 50
settlers 63
shared: desk 62; environment 4, 60–61, 68; office 39, 59, 62; spaces 7, 68, 73; start-up space 86; understanding of hot-desking policy 61; work areas, high-level 65; working environments 62, 64; workplace 68; workspaces, literature on 66
sit-stand desk (SSD) 69
smart environment (SE) 40
social: animals 13, 28; attachment 44; bonds 13–14; capital 6–7, 44, 85, 96; capital, haemorrhaging 44; change, rapid 49; changes, potential to respond to 38; cohesion 86; complexities 45; connections corroded 29; constructs 64; continuity 86; contract

redefined 96; decisions 45; engagement encouraged 70; environment, constantly changing 63; fabric 31; functions of oxytocin 92; group 31; group or organisational style 61; individuals 97; intelligence 44–45; interaction 28; mechanism 84; participation 26; PLAY 12; production of space 85; recognition 92; rejection 28; resources 15; self 84; separation 87; settings 44; skills 93; sphere, belonging to 55; stairs 70; structure 63; system 12, 60; system, office as 60
social network 7, 29, 39–40; analysis (SNA) 97–98
socio-emotional abilities 47
Spenser, E. 83
spiritual: aspects of environmental design 49; hub 43; self 84; spaces 47; views of people-environment relations 48
Stanley, R.O. 10, *11*
Stokols, D. 47–48
Stone, P.J. 65
Sullivan, L.H. 21–22, 24, 97
Sweet, F. 23
symbolic 61; aspects of environmental design 49; association, evolving 84; descriptors 90; impact of building 90; importance of self-made products 91; indicators 84; interpretations 49; meanings 48; meaning of in-between spaces 87; personalisation of space 86; products 91; use of objects in lived space 86
symbolic aspects of workspace 83; features of **48**; function of elements in 83; impact of dividers on outcomes 89; importance of 90
symbolic dimensions: required in work environment 97; in sensemaking 51–52, 53, **54**, 56, 75
symbolic emotions/sentiments 75; as part of worldview 58; reluctance to express 96; understanding of 68
symbolism 51–52, 53, 55–56, 83–86, 94, 98
symbology 51, 94
symbol (s) 83–84, 86, 90, 94

technology/technologies 4; advances in 33; augmented 39; becoming increasingly sophisticated 36; communication 1; currently available for office cooling 37; EEG 78; and emotionally intelligent workspace 41; encroachment of 24; express frustration with 58; falls short of emotional understanding of users 40; fixing 58; functional MRI (fMRI) 78; heat-generating 35; hidden in the background 39; humanity relationship with 70; imaging 78; immersive 78; *Managing the Flow of* 39; merging of telecommunications and computer 24; new, in the workplace 32, 34; organisational performance without recourse to 56; power of 2; precarious relationship with emotion 31; promised advances in productivity 35; provided all the answers 6; sorting out 15; tagging and tracking 9; tracking 30, 39–40, 97–98; traditional sensor 39; tools and methods for mapping spaces and understanding human experience 78; Wi-Fi 65; world of 58
technology building 2–3, 35; design as important 36; intelligent 3, 32, 36, 41; sentient 38
technology-laden: intelligent building 3, 32, 36; open plan environment 65
telecommunications 24; high volumes of 36
temperament 6, 10, 15, *16*, 17, 19, 21, 66, 92–94, 97–98
territorial boundaries clarified 68; boundaries established 66; claims of ownership 63; office 66; workplace, determinants of anger in 67
Thaler, R.H. 69
tick, what makes: employees in workplace 8; new type of employee 25; people 21, 25, 28; teams 1
Toffler, A. 32, 49
track 98; human behaviour 39–40; and monitor user behaviour 3
tracking: evolution of building intelligence 31, 34; information 40; technologies 9, 30, 39–40, 97–98; tools 40
trust 1, 3, 21, 98; chemical oxytocin 92–93; culture 92; raising levels of 93
trustworthy/trustworthiness 92
Turner, L. 78–79
Two-Factor Theory of Herzberg 28, 58

vagrants 63; *see also* nomad workers
Venema, T.A.G. 69–70
Vilnai-Yavetz, I. 51–52, 54

Waber, B. 30, 39
Weick, K.E. 49
Wi-Fi 37, 65; biological 92; connectivity 6; workplace 3
wireless communication and environmental control 6; systems able to communicate with one another 39; Wi-Fi communication 37
working practices 6, 8
working/work environments 39, 41; adapted to reflect identity of employees 84; aesthetic features 72; aesthetic and symbolic dimensions required for creating 97; attentional systems continually drawn to possible threats 27; corporate 86; creating 'naturalness' in 76; designers responsible for subjecting people to future shock 49; embracing aesthetic qualities of nature 75;

working/work environments *continued*
  emotional intelligence (EQ) of 47; home 3; hot-desking 4; hygiene perspective embodied in 70; influence emotional well-being 47; modern 6, 13, 19, 26, 34, 66; open plan 28; 'pull' of 14; satisfies existence needs 30; shared 62, 64; undisturbed 1
workplace/workplaces 43; achieving instrumental goals in 63; aesthetic 72, 74; affective aspects bringing about positive organisational outcomes 58; affects everyday well-being and decision-making 72; anger felt about encroachment in 67; building automation systems 35; choices 58, 65; clues, making sense of 84; comfort requirements, remedies for 37; communication in 30; contribution of people in 9; creative 74; cutting-edge 70; designers 4, 15, 58, 68; digital business 35; discipline 61; dissatisfaction 28; distractions, avoidance of 73; emotionally intelligent 47, 97; equipment providing system of meaning and purpose 73; fitting the person to 61; hierarchy of needs 26; identity 86; identity markers 86; instrumental 58, 68, 72; interaction of people in 13; IQ, attempt to measure 45; mobile workers no longer owning their own 63; modern/contemporary 3–4, 26, 63, 97–98; motivation 30; need for adaptive behaviour reduced 26–27; needs to enliven senses of employees 54; new technologies in 32, 34, 36; non-territorial 59; order, creation of 86; organic, benefits of 76; performance explained by employee relationships 40; personalise 86; professional facilities manager 35; psychological importance of objects 68; shared 68; spaces and objects in 83; strong attachment of employees to things in 66; symbolic function 83; tension from worker cooperation or lack of it 60; territorial 67; waste 60; Wi-Fi 3
workplace, emotions in: denial of 15; dysfunction created 47; interact with motivation in 8; needs of workers in 3; prevalence in 59; understanding of 9
workplace design 28; capacity to re-engage with employees 9; major innovation in 65; preoccupation with function 97; relevance of emotion of others in 4; role of coproduction, participatory design and community-based facilities management 91; science and evidence driving 56; understanding of belongingness in 28
workplace environments 45–46; ability to convey narratives and dialogues 49; create 78; emotional intelligence harnessed to support aspects of 56; emotional regulation studies in 40; inorganic 75; may enhance or hinder workers' abilities 46; natural 76; place of emotion in 15; prevalent, different from 65; user requirements in 63; without visual complexity 78
workspaces 1, 8, 21; aesthetic features of 72; affects new types of employee 25; as artefact 42, 50; creators of 15, 17; collaborative consumption of 96; as crime scene 88; cues to enable workers to feel grounded 55; dedicated 60; designers of 19, 77; efforts to coerce people to engage within 30; emotionally intelligent 2, 10, 25, 38, 41, 45, **48**, 97; emotions key to relationship with 96; environment 23; experience of infringement in 67; flexible 68; provide a motivating factor at work 19; getting the design right 62; help in satisfying higher needs 26; hygienic 78; instrumentality of 54; intelligent 45, **48**; interaction with 80; interactive environment 75; items add value to our daily routines 60; loss of everyday ownership of 63; as motivating factors 28; no fixed personal 59; people mark and protect 59; potential to provide motivation 55; provide tools and setting to get work done 72; reinvention of 9; response of employees to 27; response mood/emotions/temperament 21; -as-service 96; as set of tools 55; shared, literature about 66; solution, dominant 89; solutions, imposing normative order on 86; symbolic aspect of 83; three dimensions of 50, **54**; two-stream 63; viewed as working/not working 58; workers personalise 86
workspaces contemporary, aesthetics and symbolism necessary 98; design, instrumental outlook on 97

Zak, P.J. 92–93